David C. Billington

The Inositol Phosphates

Chemical Synthesis
and Biological Significance

© VCH Verlagsgesellschaft mbH, D-6940 Weinheim (Federal Republic of Germany), 1993

Distribution:

VCH Verlagsgesellschaft, P. O. Box 10 1161, D-6940 Weinheim (Federal Republic of Germany)

Switzerland: VCH Verlags-AG, P. O. Box, CH-4020 Basel (Switzerland)

Great Britain and Ireland: VCH Publishers (UK) Ltd., 8 Wellington Court, Wellington Street, Cambridge CB1 1HZ (Great Britain)

USA and Canada: VCH Publishers, 220 East 23rd Street, New York, NY 10010-4606 (USA)

ISBN 3-527-28152-5 (VCH Verlagsgesellschaft)   ISBN 0-89573-977-1 (VCH Publishers)

David C. Billington

The Inositol Phosphates

Chemical Synthesis
and Biological Significance

VCH   Weinheim · New York · Basel · Cambridge

Dr. David C. Billington
Institut de Recherches Servier
11, rue des Moulineaux
92150 Suresnes
France

Published jointly by
VCH Verlagsgesellschaft, Weinheim (Federal Republic of Germany)
VCH Publishers, New York, NY (USA)

Editorial Director: Dr. Don Emerson

A catalogue record for this book is available from the British Library

Die Deutsche Bibliothek – CIP-Einheitsaufnahme
Billington, David C.:
The inositol phosphates : chemical synthesis and biological
significance / David C. Billington. – 1. Aufl. – Weinheim ;
New York ; Basel ; Cambridge : VCH, 1993
    ISBN 3-527-28152-5 (Weinheim ...)
    ISBN 0-89573-977-1 (New York)

Printing: strauss offsetdruck, D-6945 Hirschberg
Bookbinding: Verlagsbuchbinderei G. Kränkl, D-6148 Heppenheim
Printed in the Federal Republic of Germany

Preface

The last 8 years have seen the study of the chemistry and biology of the inositol phosphates transformed from a quiet area into one of the largest growth areas in science today.

It is now clear that the receptor-controlled hydrolysis of inositol phospholipids, leading to intracellular second messengers, is a fundamental and widespread mechanism for the transmission of signals across cell membranes (Signal Transduction). The inositol phosphate signalling system, merely a vague hypothesis in the early 1980s, now stands alongside the classical adenylate cyclase system as a primary mechanism for the coupling of cell surface receptors to cellular responses. The chemical synthesis of the inositol phosphates, and more recently of analogues designed as tools to study the signalling cascade, is a major research theme in organic chemistry today.

The widespread occurrence of phosphate esters of *myo*-inositol has been recognised for over a century. In 1872 a highly phosphorylated substance later shown to be *myo*-inositol hexakisphosphate was isolated from seeds by Pfeffer.[1] This compound was subsequently demonstrated to be the major storage medium for phosphorus in seeds and grains, indeed *phytic acid* as it became known accounts for 60 - 90 % of the organically bound phosphorus found in seeds. This discovery led to a wide and diffuse literature concerning phytic acid, and its less highly phosphorylated sisters the inositol phosphates, in the area of food science and agriculture.

myo-Inositol	*myo*-Inositol	Phytic acid

Enzymes were discovered (phytases) which were able to hydrolyse phytic acid to less highly phosphorylated *myo*-inositol derivatives (lower phosphoric esters), thus releasing the bound phosphorus for use by the organism. Some 60 years after the original work of Pfeffer, Cason and Anderson[2] showed that these lower phosphoric esters of *myo*-inositol were also constituents of a group of phospholipids which occurred in both animal and plant tissues. These phospholipids were subsequently shown to be in a state of flux, constantly being degraded and resynthesised *via* the inositol phosphates in what came to be called a "futile cycle". How quick we were to label a system we could not understand as useless ! Nature rarely, if ever, expends hard-won metabolic energy on fruitless degradation and resynthesis of complex molecules.

The turnover of this cycle was shown[3] by Hokin and Hokin in 1953 to be stimulated by agonists such as acetylcholine, and subsequently other neurotransmitters and hormones were demonstrated to evoke an increase in the turnover of inositol phospholipids. Despite this seminal observation, little progress was forthcoming in understanding the role of the phosphoinositide (PI) cycle.

In 1980 an excellent comprehensive review of the chemistry of the inositol phosphates was published by Cosgrove,[4] and he summed up the state of the field at that time very accurately in the following words *"The lower phosphoric esters of myo-inositol do not appear to exist in other than small amounts as transient intermediates in biochemical reactions"*. This statement presumably referred to the "futile cycle" outlined above.

The seeds of our current knowledge of the central role played by these "lower esters" in cell signalling processes had however already been sown some 5 years earlier by Bob Michell.[5] In a seminal review published in 1975, he speculated that the agonist-stimulated hydrolysis of the inositol phospholipid phosphatidylinositol 4,5-bisphosphate (PtdIns) had some role to play in calcium signalling processes. This statement attracted the interest of bioscientists, including Michael Berridge [6] who was at the time working at the AFRC laboratories in Cambridge. Berridge was studying the stimulus-secretion coupling mechanisms of the insect salivary gland, and had found that the stimulation produced by the neurohormone serotonin seemed to be dependent on the mobilisation of calcium from internal cellular stores.

Phosphatidylinositol 4,5-bisphosphate   D-*myo*-Inositol 1,4,5-trisphosphate

The problem that Berridge now addressed was how the information was passed from the receptors on the cell outer surface which received the stimulatory hormone serotonin, to these internal calcium stores. His studies came to fruition in 1983 with the discovery [7] that *myo*-inositol 1,4,5-trisphosphate (1,4,5-IP$_3$), released into the cell by the agonist-stimulated hydrolysis of the cell membrane phospholipid PtdIns, acted as a second messenger and caused the mobilisation of calcium from intracellular stores.

Berridge's original observation sparked massive interest in this new signalling system, the complexity of which rapidly became apparent. Bioscientists throughout the world began to study the agonist-stimulated hydrolysis of PtdIns, and to isolate and characterise the enzymes involved in the recycling of the inositol phosphates. These investigations naturally required the availability of the various inositol phosphates proven and speculated to be intermediates in the PI cycle, and led to widespread interest in the chemical synthesis of these complex molecules.[8]

As the natural molecules became available by chemical synthesis in reasonable quantities, the synthetic emphasis has shifted in the late 1980s and early 1990s to addressing the need for analogues of these materials which could be used as tools to probe the biochemical pathways involved. This has led to a renaissance in the chemical synthesis of the inositol phosphates [8] and their analogues, most of which has been published in the last 4 years. Synthetic interest is still very high in this area, with the prospect that manipulation of the signalling system could lead to valuable new therapies for a number of disease states, adding fuel to the fire.

The GPI anchor of *Trypanosoma brucei*

The biochemistry of the inositol phosphate signalling system is still the subject of intense study, and while agreement on the broad outline of the events involved has been reached, there is still considerable debate as to the fine details of the processes involved.[9] In addition, a number of other important roles have come to light for *myo*-inositol and the inositol phosphates. It is now clear for example that inositol phosphate esters play a key role in the linkage of a large number (30+) of surface glycoproteins to cell membranes [10] *via* the glycophosphatidyl inositol (GPI) anchor molecule, and that a pair of phosphooligo-saccharides containing an inositol phosphate group, of strikingly similar structure to the GPI may have insulin mimetic activity .[11] Indeed either or both of these phosphooligosaccharides may be the intracellular second messengers of insulin action ,[12] perhaps derived from the GPI.

For the above reasons the brief description of the biochemical importance of the inositol phosphates given in this book reflects a personal perspective of the current state of an evolving investigation.

As a primary goal of this book I have attempted to collect all of the syntheses of the inositol phosphates reported up to June 1991. I have also included analogues of the inositol phosphates which have been synthesised recently as specific tools for biochemical investigations.

A limited coverage of syntheses appearing after June 1991 is presented in Chapter 9, Late Entries.

References

1)   E.Pfeffer, *Pringsheims Jb.Wis.Bot.*, 1872, **8**, 429 ; 475.

2)   J.Cason and R.J.Anderson, *J.Biol.Chem.*, 1938, **126**, 527.

3)   M.R.Hokin and L.E.Hokin, *J.Biol.Chem*, 1953, *203*, 967; L.E.Hokin, *Ann.Rev.Biochem.*, 1985, **54**, 205.

4)   D.J.Cosgrove, "Inositol Phosphates, Their Chemistry Biochemistry and Physiology", 1980, Elsevier, Oxford.

5)   R.H.Michell, *Biochim.Biophys.Acta.*, 1975, **415**, 81.

6)   M.J.Berridge, *J.Am.Med.Assoc.*, 1989, **262**, 1834.

7)   H.Streb, R.F.Irvine, M.J.Berridge and I.Schulz, *Nature* (London), 1983, **306**, 67; M.J.Berridge and R.F.Irvine, *Nature* (London), 1984, **312**, 315.

8)   D.C.Billington, *Chem.Soc.Rev.*, 1989, **18**, 83; B.V.L.Potter, *Nat.Prod.Rep.*, 1990, **7**, 1.

9)   R.H.Michell, A.H.Drummond and C.P.Downes, "Inositol Lipids and Cell Signalling", 1989, Academic Press, London; S.R.Nahorski, C.I.Ragan and R.A.J.Challiss, *Trends.Pharmacol.Sci.*, 1991, **12**, 297; S.B.Shears, *Pharmac.Ther.*, 1991, **49**, 79; A.A.Abdel-Latif, *Cellular Signalling*, 1991, **3**, 371; C.P.Downes and A.N.Carter, *Cellular Signalling*, 1991, **3**, 501.

10)   M.A.J.Ferguson and A.F.Williams, *Ann.Rev.Biochem.*, 1988, **57**, 285.

11)   A.R.Saltiel and P.Cuatrecasas, *Proc.Nat.Acad.Sci.USA*, 1986, **83**, 5793.

12)   M.G.Low and A.R.Saltiel, *Science*, 1988, **239**, 268.

CONTENTS

Abbreviations

AC	Adenylate cyclase
cAMP	Cyclic adenosine monophosphate
ATP	Adenosine triphosphate
Bn	Benzyl ; $CH_2C_6H_5$
Bzl	Benzoyl ; COC_6H_5
C-1, C-2, *etc*	Numbering of carbon atoms in the *myo*-inositol ring, as below :

Cam	Camphanate ester
mCPBA	*meta*-Chloroperbenzoic acid
D	Absolute configuration related to D-glyceraldehyde
DAG	Diacylglycerol
DDEO	

DMF	Dimethylformamide
DMSO	Dimethylsulphoxide
DNA	Deoxyribonucleic acid
DPCC	Diphenylchlorophosphate
GDP	Guanosine diphosphate
G-6-P	D-Glucose-6-phosphate
GPI	Glycophosphatidyl inositol
G-Protein	Guanosine nucleotide-binding protein
GTP	Guanosine triphosphate
HI	Hydroiodic acid

HPLC	High pressure (or performance) liquid chromatography
IMPase	Inositol monophosphatase
Inositol	*myo*-Inositol
I-1-P	D-Inositol 1-phosphate
I-3-P	D-Inositol 3-phosphate (L-inositol 1-phosphate)
I-4-P	D-Inositol 4-phosphate
1,3-IP$_2$	Inositol 1,3-bisphosphate
1,4-IP$_2$	D-Inositol 1,4-bisphosphate
3,4-IP$_2$	D-Inositol 3,4-bisphosphate
1,3,4-IP$_3$	D-Inositol 1,3,4-trisphosphate
1,4,5-IP$_3$	D-Inositol 1,4,5-trisphosphate
1,3,4,5-IP$_4$	D-Inositol 1,3,4,5-tetrakisphosphate
IUPAC	International Union of Pure and Applied Chemistry
L	Absolute configuration related to L-glyceraldehyde
MEM	Methoxyethoxymethyl ; $CH_2OCH_2CH_2OCH_3$
MOM	Methoxymethyl ; CH_2OCH_3
PI	Phosphatidyl inositol
PIP	Phosphatidyl inositol 4-phosphate
PIP$_2$	Phosphatidyl inositol 4,5-bisphosphate
PMB	*para*-Methoxybenzyl ; $CH_2C_6H_4OCH_3$
Pr	*n*-Propyl ; $CH_2CH_2CH_3$
rac	Racemic
s	Substrate concentration (in an enzymic reaction)
TBDMS	*tert*-Butyl dimethylsilyl
TBPP	Tetrabenzylpyrophosphate
THF	Tetrahydrofuran
THP	Tetrahydropyranyl
TIPS	1,3-Dichloro-1,1,3,3-tetraisopropyldisiloxane
v	Velocity (of an enzymic reaction)

Chapter 1

The Inositols

1.1 Discovery

In 1850 an optically inactive cyclohexane hexol was isolated from heart muscle by Scherer,[1] and he named this substance " inosit " from the Greek for muscle or sinew, *Fig 1-1*. This name came to be used as a generic term for cyclohexane hexols, as these compounds were discovered during subsequent years, with the suffix " ol " added in English and French usage. It became customary to refer to Scherer's isomer as *meso*-inositol, presumably by analogy with the optically inactive *meso*-tartaric acid. This name was in common usage for many years, despite the fact that there are, as we shall see, six other stereoisomeric inositols to which the prefix *meso*- is equally applicable.

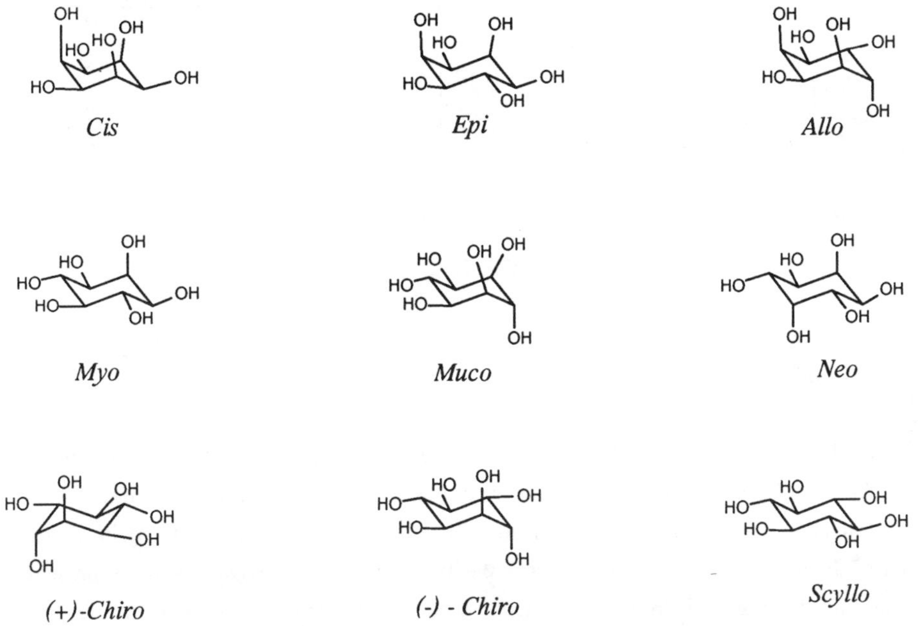

Fig 1 - 1

The fact that there are nine possible stereoisomeric inositols was recognised as long ago as 1894 by Bouveault,[2] and their structures are given in *Fig 1-1*, together with their trivial names.[3] Of these nine isomers, seven are optically inactive *meso* compounds (*allo, cis, epi, muco, myo, neo* and *scyllo*), that is to say they have an internal plane or axis of symmetry, and thus cannot exist as non-superimposable mirror images. The remaining two isomers have no such symmetry and form an enantiomeric pair [(+) and (-) *chiro*].

The inositols (cyclohexanehexols) are members of the group of "carbocyclic carbohydrates" called cyclitols.[3] This class of compounds includes cyclopentanes , cyclohexanes, and some larger ring compounds, which have multiple (>2) hydroxyl groups present . The unsaturated derivatives of these ring systems, as well as compounds related to them by O-substitution, by C-substitution, and in some cases by replacement of hydroxyl groups with other substituents, are also included in this general class. Some representative structures of six-membered cyclitols are given in *Fig 1-2*.

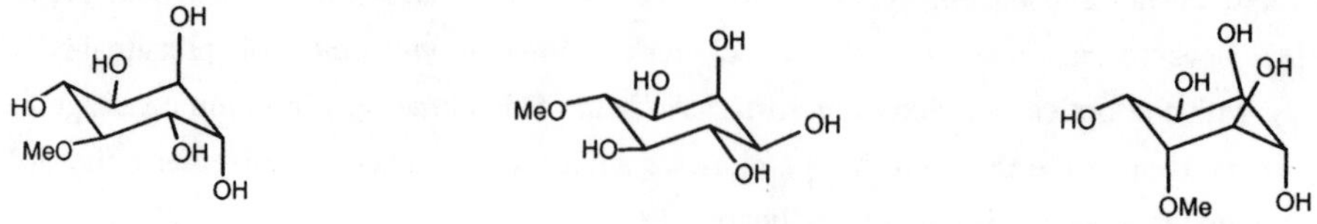

(+) - Pinitol

(+) - Ononitol

(-) - Quebrachitol

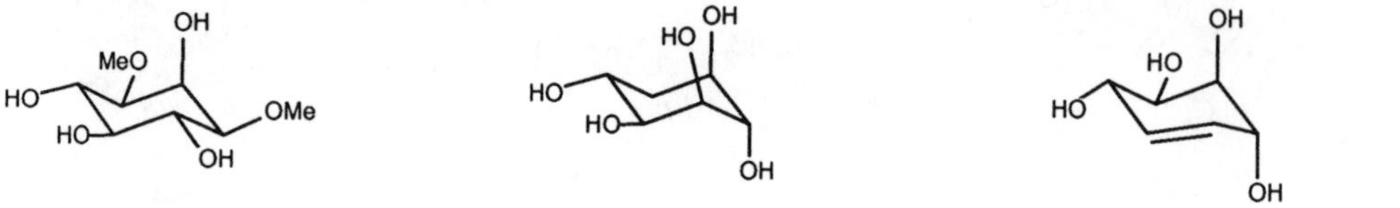

Dambonitol

(+) - Quercitol

Conduritol A

Fig 1 - 2

1.2 *myo*-Inositol

1.2.1 Occurrence

The use of the prefix *myo*- to designate the isomer first isolated by Scherer was adopted in 1951, following a suggestion by Fletcher *et al.*[4] *Myo*-inositol is the most common inositol isomer in nature, and appears to be present in either free or combined form in all living species.[5] In animals and microorganisms the major portion of *myo*-inositol is present as inositol phospholipids which are components of cell membranes, and as its hexakisphosphate, where it acts as the major phosphorus source in seeds. *Myo*-inositol is commercially prepared from corn-steep liquor in large quantities and is sold by most chemical suppliers in high purity in kilogram quantities. Unknown compounds subsequently identified as *myo*-inositol have been isolated from many sources, and this has led to a whole series of names for this cyclitol appearing in the literature, including : cyclohexanehexol, cyclohexitol, meat sugar, inosite, mesoinosite, phaseomannite, dambose, nucite, bios I, rat anti-spectacled eye factor and mouse anti-alopecia factor.[6] As we shall only be concerned with the *myo*-isomer in this book, "inositol" will be used throughout the text, and refers to the *myo*-inositol isomer unless otherwise indicated. The prefix *myo*- will only be inserted in cases where clarification is necessary, for example when a change of configuration has occurred in a synthetic operation.

1.2.2 Stereochemistry and Numbering

The naming and numbering of the inositols, and of cyclitols in general, presents many complex problems. This resulted in part from the adoption of a number of nomenclature schemes as the field matured and expanded, and can lead to considerable confusion, particularly in interpretation of the older literature. Above all, the inherent chemical complexity of the inositols leads to a multitude of problems, regardless of the most carefully formulated nomenclature rules. Nomenclature of the inositols, and of cyclitols in general has been the subject of recent reviews and IUPAC recommendations. [5,7,8,9] In particular, Cosgrove's treatment of the general nomenclature of the inositols,[5] and the introduction to the problems of *myo*-inositol stereochemistry and nomenclature published by Parthasarathy and Eisenberg [8] are recommended to the interested reader. Here, the general rules for nomenclature will not be reproduced to avoid unnecessary confusion, rather only the requirements for nomenclature in the *myo*-inositol case will be outlined.

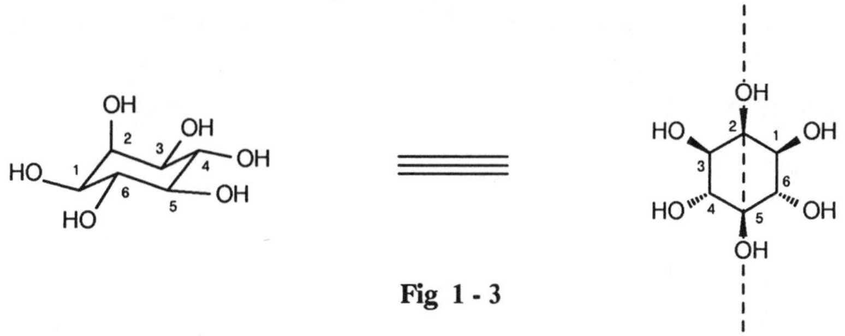

Fig 1 - 3

In what is by far the most stable conformation in solution, *myo*-inositol has a single axial hydroxyl group (numbered by convention as position 2), and five equatorial hydroxyl groups, *Fig 1-3*. This leads to a plane of symmetry which runs through C-2 and C-5, resulting in the *meso*- stereochemistry alluded to above. In order to emphasise the symmetry of the system, the two-dimensional planar representation of inositol shown in *Fig 1-3* will be used throughout the book. Incorporation of a substituent at C-2 or C-5 of *myo*-inositol leads to an optically inactive *meso*- compound, as the plane of symmetry is retained. By contrast, incorporation of a substituent at C-1 (enantiotopic to C-3) and / or C-4 (enantiotopic to C-6) leads to a pair of enantiomers (plane of symmetry lost on substitution). Thus, inositol 1-phosphate exists as a pair of enantiomers (1) and (2), inositol 2-phosphate (3) and inositol 1,3- bisphosphate (4) are both *meso* compounds, and inositol 1,4,5-trisphosphate exists as a pair of enantiomers (5) and (6), *Fig 1-4*.

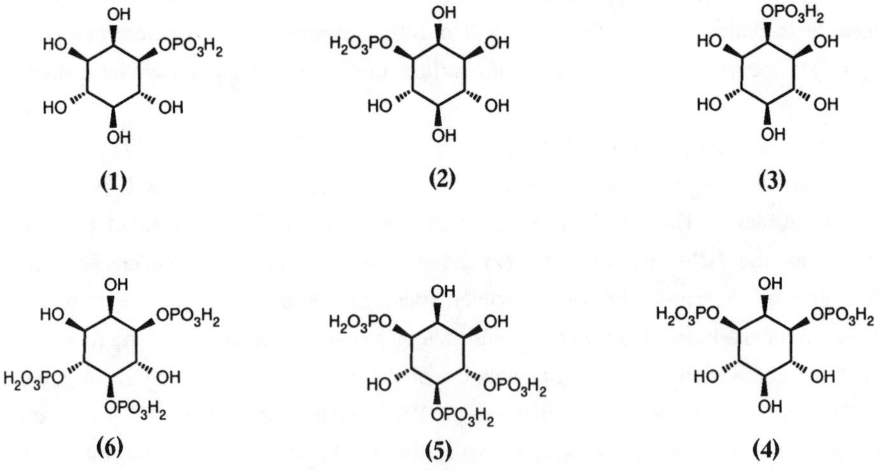

Fig 1 - 4

The enantiomers (1) and (2) exemplify the fact that, in order to obtain the lowest number count, it is equally correct to number the inositol carbons from C-2 in the representation shown either clockwise or anticlockwise. By convention, if anticlockwise numbering leads to the lowest number count, then the derivative is assigned the prefix D-, and clockwise numbering leads to the L- prefix, *Fig 1-5*. For consistency in presentation, the inositol ring

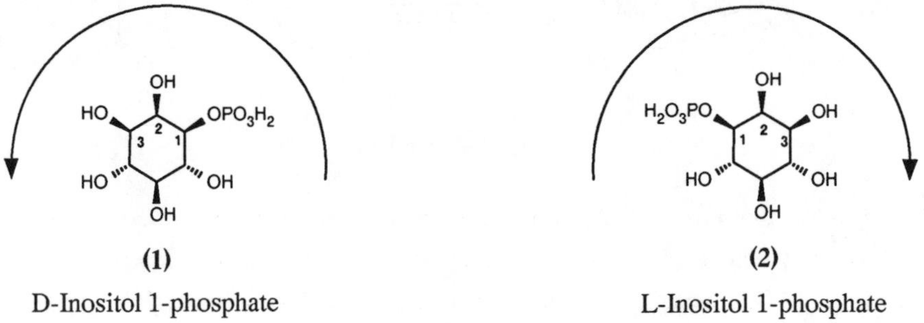

(1)   (2)

D-Inositol 1-phosphate   L-Inositol 1-phosphate

Fig 1 - 5

will be drawn and numbered as the D- isomer as far as possible, *ie* carbon atoms will be numbered anticlockwise with C-1 at 2 o'clock throughout the book. In structures where this is not the numbering convention adopted, the ring carbons will be numbered.

In addition to nomenclature problems, numbering of inositol derivatives poses problems, due in part to the symmetry characteristics outlined above, and strict adherence to the IUPAC rules can cause considerable confusion when following a given inositol derivative through a metabolic pathway. This has led to a new proposal from IUPAC,[9] which suggests that, as all of the naturally occurring inositol phosphates are of the D-configuration, *they may all be numbered for biological purposes as the D-inositol derivatives,* rather than by strict chemical sequence rules,as this may lead to a clarification in biochemical pathway interpretation. This

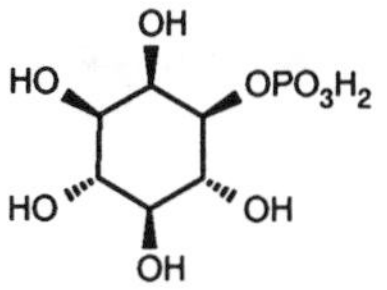

(1)	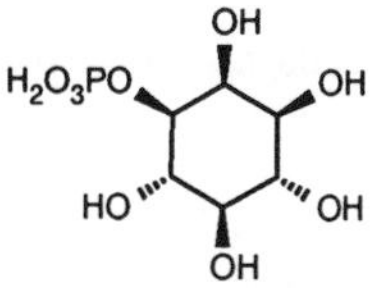	(2)
D-Inositol 1-phosphate	——— *Normal* ———	L-Inositol 1-phosphate
D-Inositol 1-phosphate	——— *Alternative* ———	D-Inositol 3-phosphate

Fig 1 - 6

leads to the two enantiomers of inositol 1-phosphate (1) and (2) being named as D-inositol 1-phosphate and D-inositol 3-phosphate, rather than as D- and L- inositol 1-phosphate, *Fig 1-6.* The advantage of this convention is that it may make it rather easier to follow the metabolic fate of the inositol phosphate isomers at times.

An example of the use of this convention is given in *Fig 1-7*, where the enzymic hydrolysis of the D-*myo*-inositol 3,4-bisphosphate shown (7) gives the monophosphate L-*myo*-inositol 1-phosphate (2). The use of the alternative IUPAC convention leads to the name D-*myo*-inositol 3-phosphate for this product (2), which makes it somewhat easier to follow the metabolic process occurring(*ie* a 3,4-bisphosphate is transformed into a 3-phosphate *via* loss of its 4-phosphate group). This numbering / nomenclature convention will only be used in parallel with the normal chemical system in the biochemical discussions.

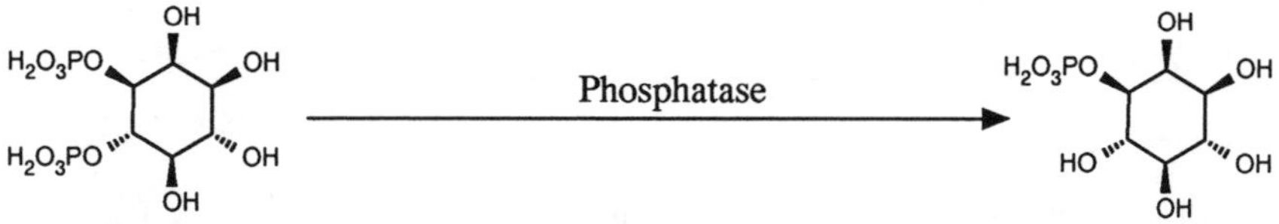

Phosphatase

(7)   (2)

Fig 1 - 7

The debate over the naming and numbering of inositol derivatives is still ongoing, and more proposals can be expected in the future.

As outlined above the C-1 at 2 o'clock / anticlockwise representation will be used as much as possible in the chemistry discussions, and any examples which require clockwise numbering to generate the correct chemical name will be indicated by numbering of the ring atoms in the schemes. This is necessary occasionally for clarity in presentation, as the symmetry of the inositol system has been exploited in a number of total syntheses. It should be borne in mind that many of the inositol derivatives in the schemes are racemic mixtures, although only a single enantiomer is drawn. To distinguish racemic syntheses from chiral syntheses, the starting materials and final products will be labelled as racemic (*rac*) or optically active [D or L ; (+) or (-)].

1.2.3 Biosynthesis of myo-Inositol

Due to the widespread occurrence of inositol in plants, most inositol in humans is of dietary origin. *De novo* biosynthesis does occur however in animals and humans *via* the isomerisation of D-glucose 6-phosphate to L-inositol 1-phosphate. The enzyme responsible for this interesting reaction, L-*myo*-inositol 1-phosphate synthase, is very abundant in mammalian testis and brain and has been purified from bovine and rat-derived tissues. The enzyme from yeast has been purified to homogeneity, and its gene regulation studied in detail. The enzyme has attracted much mechanistic interest due to the coupled stereospecific ring closure and inosose reduction which it performs. The overall process is depicted in *Fig 1-8*, and has been recently reviewed .[10,11,12]

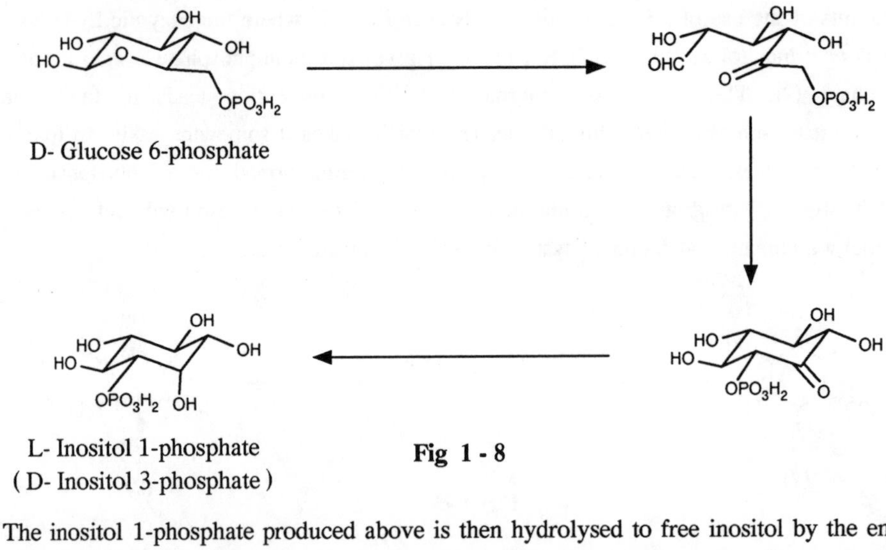

D- Glucose 6-phosphate

L- Inositol 1-phosphate
(D- Inositol 3-phosphate)

Fig 1 - 8

The inositol 1-phosphate produced above is then hydrolysed to free inositol by the enzyme *myo*-inositol 1-phosphatase, and thus the overall pathway is as shown, *Fig 1-9*.

The existence of this pathway in brain presumably arises from the fact that inositol does not cross the blood-brain barrier, except by a specific low capacity uptake system, which is saturated under normal conditions. *Myo*-inositol 1-phosphatase has recently been shown to hydrolyse both enantiomers of inositol 1-phosphate, and both enantiomers of inositol

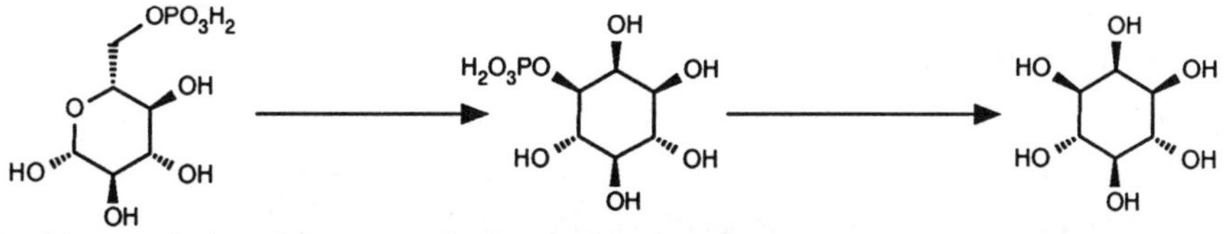

D- Glucose 6-phosphate   L- Inositol 1-phosphate   *myo*- Inositol
(D- Inositol 3-phosphate)

Fig 1 - 9

4-phosphate.[13] This seeming lack of specificity probably reflects the symmetry inherent in the inositol phosphates, coupled with a three- or four-point recognition requirement by the enzyme, and has led to the proposal that inositol monophosphatase would be a more fitting name for this enzyme.

Inositol monophosphatase is of great current interest due to the fact that it is inhibited in an *uncompetitive* manner by lithium ions, and that levels of inositol in the brains of lithium-treated rats are reduced compared with controls.[14,15] As treatment with lithium salts is currently the only effective therapy for manic depression, this has led to speculation that an alternative inhibitor of the monophosphatase could provide a replacement therapy for this debilitating disease. This aspect of inositol phosphate metabolism, and its relationship to receptor signalling and the PI cycle in general, is covered in more detail in Chapter 2.

References

1)   J.Scherer, *Ann.*, 1850, 322.

2)   L.Bouveault, *Bull.Chim.Soc.Fr.*, 1894, **11**, 144.

3)   T.Posternak, "The Cyclitols", 1965, Holden-Day, San Fransisco.

4)   H.G.Fletcher, L.Anderson and H.Lardy, *J.Org.Chem.*, 1951, **16**, 1238.

5)   D.J.Cosgrove, "Inositol Phosphates, Their Chemistry Biochemistry and Physiology",1980, Elsevier, Oxford.

6)   The Merck Index, 11th edition, Merck and Co Inc, Rahway, New Jersey.

7)   IUPAC, "The 1967 IUPAC Tentative Rules For Cyclitols", *IUPAC Information Bull.*, 1968, **32**, 51.

8)   R.Parthasarathy and F.Eisenberg Jr., *Biochem.J.*, 1986, **235**, 313.

9)   IUPAC, *Biochem.J.*, 1989, **258**, 1.

10)   B.V.L.Potter, Chapter 11.4 in "Comprehensive Medicinal Chemistry", 1990, Pergamon Press, Oxford.

11)   F.A.Loewus, *Recent Adv.Phytochem.*, 1974, **8**, 179.

12)   F.A.Loewus and M.W.Loewus, *Ann.Rev.Plant Physiol.*, 1983, **34**, 137.

13)   N.S.Gee, C.I.Ragan, K.J.Watling, S.Aspley, R.G.Jackson, G.G.Reid, D.Gani and J.K.Shute, *Biochem.J.*, 1988, **249**, 883.

14)   W.R.Sherman, in " Inositol Lipids in Cell Signalling", 1989, R.H.Michell, A.H.Drummond and C.P.Downes eds, Academic Press, London, p 39.

15)   A.H.Drummond, *Trends Pharmacol.Sci.*, 1987, **8**, 129.

Chapter 2

Cell-Cell Signalling and the Inositol Phosphates

2.1 Introduction to Cell-Cell Signalling

Complex organisms rely on communication between individual cells in order to maintain life functions.[1] It is now universally accepted that this communication involves the use of distinct chemicals to pass messages from one cell to another, and these chemical messengers may be classified by their function, as neurotransmitters, hormones, growth factors, *etc*. With the exception of steroid hormones,[2] which diffuse freely across cell membranes and are recognised by proteins inside the target cell, the chemical messengers released by cells are recognised by proteins termed receptors on the outer surfaces of the target cells membrane. These receptors must then process the incoming signal from the outer surface of the cell membrane to produce a signal inside the target cell.[3] This conversion of an extracellular chemical signal into an intracellular signal by passing information across the cell membrane is called signal transduction, and receptors use a variety of mechanisms to achieve this transduction. Once the message has been passed across the cell membrane, the target cell then responds to this intracellular signal, and thus the stimulation of one cell leads, *via* extracellular chemical messengers, receptors, signal transduction, and the production of an intracellular signal, to a response in a second physically distant cell, *Fig 2-1*.

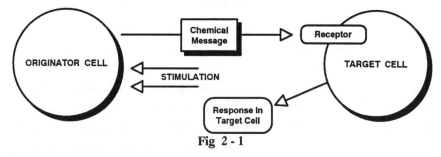

Fig 2 - 1

2.2 Types of Receptors

One class of cell surface receptors either contains, or is closely linked to, an ion channel which spans the cell membrane, *Fig 2-2*.[3,4] Stimulation of the receptor leads to an alteration in the ability of the channel to allow ions to pass into or out of the target cell. Thus the concentration of a given ionic species changes inside the cell, and this is perceived as an internal signal. This type of receptor is widely used by cells sensitive to neurotransmitters.

A second class of receptors are actually enzymes, called tyrosine kinases, which are embedded in the cell membrane. The tyrosine kinase receptors have sites on the outer surface which recognise agonists, and active sites on the inside of the cell membrane. The binding of

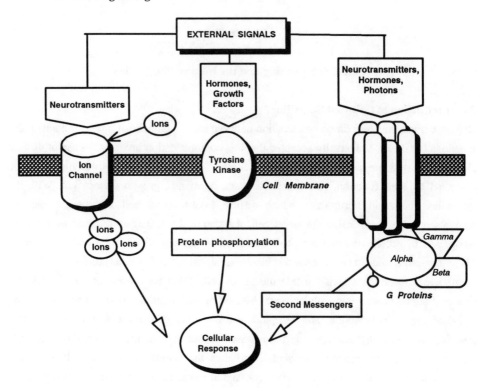

Fig 2 - 2

extracellular agonists to these enzymes modulates the activity of the enzymic species, causing the phosphorylation of tyrosine residues on target proteins inside the cell. This type of signal transduction is used by many growth factors and hormones, including insulin.[5]

The third class of receptors has no intrinsic activity as either ion channel control systems or enzymes. These receptors are characterised by a set of seven transmembrane helical domains and are coupled *via* a class of proteins known as guanosine nucleotide-binding proteins, or G-proteins,[6] to the intracellular enzymes or ion channels through which they evoke their responses. Receptors which use this type of coupled receptor may be found for all kinds of intercellular signals.

2.3 G-Protein Coupled Receptors

The G-proteins involved in cell-cell signalling belong to a super-family of guanosine nucleotide-binding proteins which includes cytoskeletal proteins like tubulin, soluble proteins, and low molecular weight GTP binding proteins like the *ras* p21 protooncogenes. The G-proteins associated with receptor signalling all exist as heterotrimeric structures, containing one alpha, one beta, and one gamma subunit, and are found associated with the internal surface of the cell membrane.[7] The latter two subunits are tightly bound in a non-covalent beta/gamma complex. The mechanism by which G-proteins act as signal transducers is outlined in *Fig 2-3*.[8] On the left the receptor is not occupied by an agonist and is in the off position. The G-protein trimeric complex is in its inactive form, with guanosine

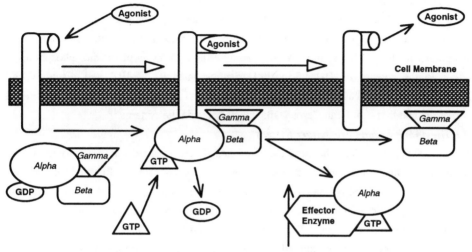

Fig 2 - 3

diphosphate (GDP) bound to the alpha subunit, and is associated with the inner surface of the cell membrane. Binding of the natural ligand, or other agonist molecule to the receptor causes a conformational change in the receptor protein, resulting in the formation of a ternary complex between the receptor protein and the GDP-protein complex. This complex then instantly releases GDP and binds guanosine triphosphate (GTP), and the GTP-protein complex dissociates from the receptor. The activated alpha subunit with GTP bound to it then dissociates from the beta/gamma dimer and it is this alpha-GTP subunit which binds to the effector system to regulate either the opening and closing of an ion channel, or the activity of the enzyme to which the receptor is coupled. This binding association affects the enzyme's catalytic activity, either increasing or decreasing the rate at which it forms active second messenger molecules from inactive substrates within the cell. In time the bound GTP is hydrolysed to GDP by the inherent GTPase activity of the alpha subunit, and the resulting alpha-GDP complex dissociates from the effector thus re-setting the switch. The alpha-GDP complex then re-associates with the beta/gamma complex to return the system to its original state. In terms of structure the alpha subunit contains both the GTP binding site, and the catalytic site for GTPase activity. At least twenty distinct alpha subunits have been identified to date, with *ca* 40% sequence homology being preserved across the range of proteins. At least two distinct genes exist for both the beta and gamma subunits, leading to a diverse array of heterotrimeric G-proteins. Our knowledge of the structure and function of the G-proteins has been expanding at an enormous rate over the last 10 years or so, and many review articles are available on this topic.[6,7,8,9,10,11]

2.4 Second Messengers

G-protein coupled receptors that act *via* effector enzymes to produce second messengers can be divided into two types, those which stimulate the production of cyclic 3',5'-adenosine monophosphate (cAMP), and those which promote the hydrolysis of inositol phospholipids leading to the release of D-1,4,5-inositol trisphosphate (1,4,5-IP$_3$) and diacylglycerol (DAG).[1,4] There is increasing evidence that these second messenger systems do not operate in isolation, and that significant "cross-talk" exists between the systems.[12]

2.4.1 The cAMP Second Messenger System

Receptors which stimulate the production of cAMP are coupled *via* a G-protein to the enzyme adenylate cyclase, which converts adenosine triphosphate (ATP) to cAMP. cAMP was identified as the active second messenger in this system in 1959 , and it acts by stimulating a specific protein kinase, which in turn phosphorylates target proteins in the cell, leading to the overall response.[13] The cAMP system has been found to regulate intracellular reactions in all procaryotic and animal cells examined to date. Examples of receptors which

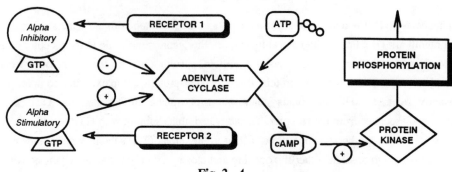

Fig 2 - 4

use the stimulation of adenylate cyclase as an effector system include the *beta*$_1$-adrenoceptor and the dopaminergic D$_1$ receptor. The activation of other receptors, the *alpha*$_2$-adrenoceptor for example, can slow down the formation of cAMP *via* production of an alpha-GTP complex which inhibits adenylate cyclase activity. Indeed in some cells both inhibitory and excitatory receptors seem to exist, and the overall level of cAMP therefore depends on the final balance between the activity of the two types of receptors. The overall pathway is outlined in *Fig 2-4*.

2.4.2 The Inositol Phospholipid Second Messenger System

A second more recently discovered cell-cell signalling system uses increases in the concentration of calcium in the target cell to evoke cellular responses.[14] In resting cells the concentration of calcium inside the cell is very low, *ca* 10^{-7} molar, compared to its normal concentration in the extracellular fluid of *ca* 10^{-3} molar. This gradient is maintained by two general mechanisms, simple pumps which pump calcium ions out of the cell, and specific calcium uptake systems which sequester calcium inside storage areas in the cell. Many cellular processes and enzymic systems are highly calcium dependent, and receptor mediated increases in cell calcium concentrations from *ca* 10^{-7} to *ca* 5×10^{-6} molar constitute a ubiquitous cell-cell signalling system.

The examination of the relationship between the turnover of inositol phospholipids and intracellular calcium levels began in the 1950s with an observation made by the husband and wife team Hokin and Hokin, who found that stimulation of the muscarinic cholinergic receptor by cholinergic agonists such as acetylcholine selectively increased the incorporation of ^{32}P into two minor, inositol containing, cell membrane lipids.[15] Almost 25 years passed before Michell noted that receptors which stimulated phosphoinositide turnover also activated calcium dependent processes inside the cell, which led him to propose that receptor-stimulated turnover of inositol phospholipids resulted in increases in intracellular calcium levels, and could therefore constitute a new cell-cell signalling pathway.[16] It is now clear that a relatively minor cell membrane lipid phosphatidylinositol (PI) is sequentially phosphorylated by specific kinases in the cell, first to phosphatidylinositol-4-phosphate, (PIP) and then to phosphatidylinositol-4,5-bisphosphate (PIP$_2$), *Fig 2-5*. This phospholipid PIP$_2$ is

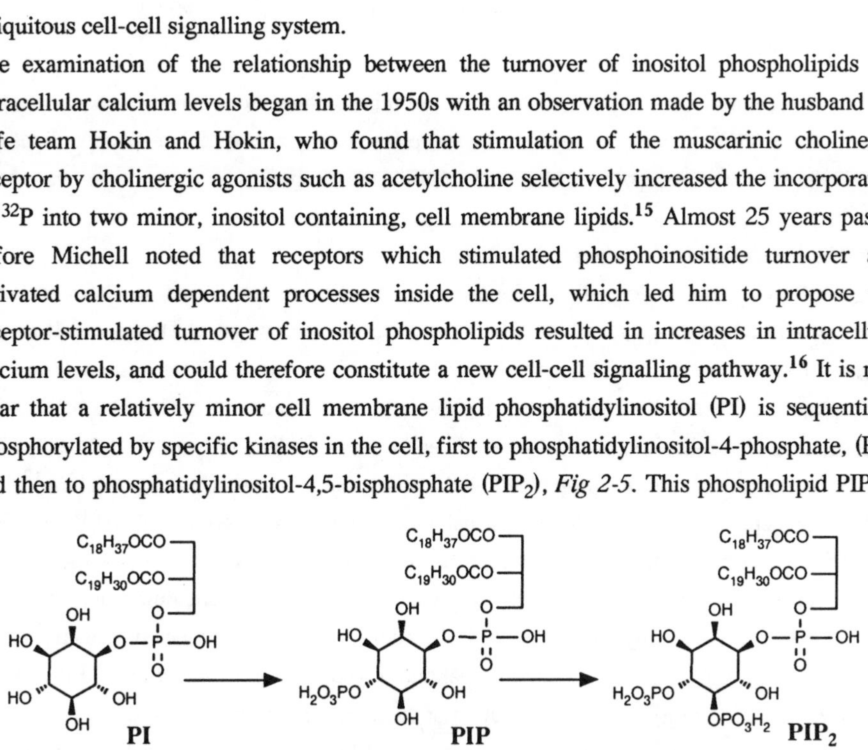

Fig 2 - 5

the substrate for a hydrolytic enzyme, phospholipase C, which is under the control of cell surface receptors *via* the intermediacy of a G-protein system, as seen above for the enzyme adenylate cyclase. Activated phospholipase C cleaves PIP$_2$ to give water-soluble D-1,4,5-inositol trisphosphate (1,4,5-IP$_3$) and lipid-soluble diacylglycerol (DAG), which remains in the plane of the cell membrane, *Fig 2-6*. Both of these molecules act as second messengers inside the target cell, 1,4,5-IP$_3$ acting to increase calcium concentration, whilst DAG binds to and activates a specific protein kinase, protein kinase C, promoting phosphorylation of target proteins in the cell.[17] In addition, DAG can be further metabolised to produce arachidonic acid, a starting point for the synthesis of the prostaglandins. A large

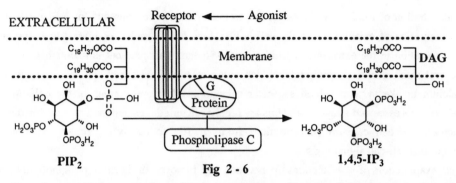

Fig 2 - 6

number of receptors have now been shown to use this mechanism of signal transduction, and as with the cAMP system, in addition to receptors which can stimulate phospholipase C hydrolysis of PIP_2, receptors which can inhibit the enzyme's activity have also been identified.

2.4.3 The Inositol Polyphosphates

In the following discussion all of the inositol phosphates are drawn in their naturally occurring absolute configurations, as the D-isomers, and are named as D-isomers. It should be remembered however that due to the symmetry of the *myo*-inositol system D-inositol 3-phosphate is identical to L-inositol 1-phosphate, see Chapter 1. The D-1,4,5-inositol trisphosphate ($1,4,5$-IP_3) produced by the receptor controlled hydrolysis of PIP_2 acts upon specific receptors inside the target cell and causes the release of stored calcium.[18] The $1,4,5$-IP_3 receptors which mediate this calcium release have been isolated,[19] purified, cloned,[20] sequenced,[21] and reconstituted[22] from a variety of tissues. The location of the calcium stores themselves is probably the rough endoplasmic reticulum, but the exact location and the possibility of cell to cell variations is still the subject of debate.[23] This release of calcium causes the intracellular calcium concentration to increase, and thus leads to the overall cellular response by modulation of calcium dependent processes.

The $1,4,5$-IP_3 calcium mobilising signal is terminated by the metabolism of $1,4,5$-IP_3 by one of two routes, *Fig 2-7*. In the first route a specific 5-phosphatase removes the 5-phosphate group of $1,4,5$-IP_3, giving D-inositol 1,4-bisphosphate ($1,4$-IP_2) which is not active as a calcium mobilising agent and thus the signal is terminated.[24,25] Further metabolism of $1,4$-IP_2 by hydrolysis of the 1-phosphate group then gives D-inositol 4-phosphate (I-4-P), which is in turn hydrolysed to free inositol by the enzyme inositol monophosphatase (IMPase). The second pathway for the metabolism of $1,4,5$-IP_3 involves phosphorylation of the 3-position by a specific 3-kinase, giving D-1,3,4,5-inositol tetrakisphosphate ($1,3,4,5$-IP_4). This tetrakisphosphate shows a variety of effects in stimulated cells, and at present its role *in vivo* is not at all clear.[26] One thing that is clear however is that in the presence of $1,4,5$-IP_3, and only in its presence, $1,3,4,5$-IP_4 can stimulate the entry of calcium ions into cells. There

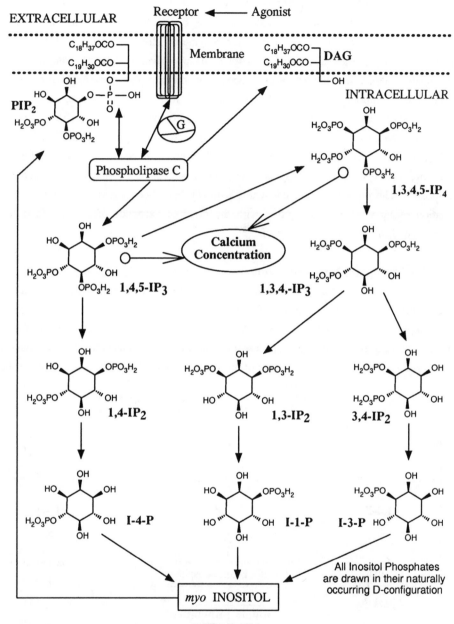

Fig 2 - 7

seems to be a complex synergysm between 1,4,5-IP$_3$ and 1,3,4,5-IP$_4$, their respective receptors, and the overall calcium balance of stimulated cells.[27] A functional model has been proposed to account for these observations.[28] The first step in the metabolism of 1,3,4,5-IP$_4$

involves the action of the same 5-phosphatase which hydrolyses 1,4,5-IP$_3$, in this case giving rise to an isomeric trisphosphate, D-1,3,4 inositol trisphosphate (1,3,4-IP$_3$), which appears to be biologically inactive. The action of the same enzyme which hydrolyses 1,4-IP$_2$ to I-4-P on this new trisphosphate gives rise to D-inositol 3,4-bisphosphate (3,4-IP$_2$) by removal of the 1-phosphate group, whilst the action of a second enzyme, a 4-phosphatase, gives rise to the isomeric *meso* compound inositol 1,3-bisphosphate (1,3-IP$_2$). This 4-phosphatase also attacks 3,4-IP$_2$, removing the 4-phosphate group to give D-inositol 3-phosphate (I-3-P)(*ie* L-inositol 1-phosphate see Chapter 1). The *meso* 1,3-IP$_2$ is metabolised by a specific 3-phosphatase which gives rise to D-inositol 1-phosphate (I-1-P). Both I-1-P and I-3-P are hydrolysed to free inositol by the same monophosphatase enzyme which converts D-inositol 4-phosphate to free inositol. The enzymes involved in this pathway have been studied in detail, and the regulation of these enzymes, together with the possible existence of isoforms and their compartmentalisation has been reviewed.[24,25]

2.5 Inositol Monophosphatase and Lithium

From the above outline it can be seen that metabolism of the PIP$_2$ derived calcium mobilising signal 1,4,5-IP$_3$ by either pathway finally gives rise to three isomeric D-inositol monophosphates, I-4-P, I-1-P and I-3-P. A single enzyme, IMPase, is responsible for the hydrolysis of all of these compounds to free inositol, which is then used for the resynthesis of PIP$_2$, *Fig 2-8*. This recycling of inositol to provide more PIP$_2$ for receptor controlled

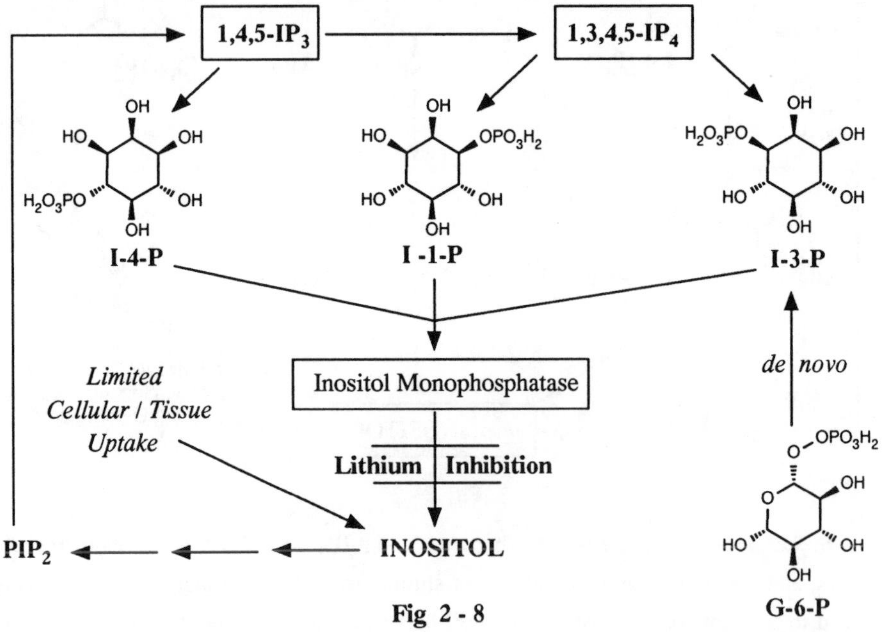

Fig 2 - 8

generation of 1,4,5-IP$_3$ is particularly important in the brain, as inositol does not readily cross the blood-brain barrier, and is normally only transported into the brain by an active transport system of very limited capacity.[29] Limited *de novo* synthesis of inositol from D-glucose 6-phosphate (G-6-P) occurs in the brain, *via* conversion of G-6-P into I-3-P, which is then hydrolysed to free inositol by IMPase, see Chapter 1. Thus hydrolysis of the various inositol monophosphates by the enzyme IMPase represents a crucial step in the production of free inositol, and thus of both PIP$_2$ and 1,4,5-IP$_3$ in stimulated cells, *Fig 2-8*, and inhibition of this enzyme might be expected to have profound effects on the rate of cell signalling.

Both IMPase, and the inositol polyphosphate 1-phosphatase which hydrolyses 1,4-IP$_2$ and 1,3,4-IP$_3$ are inhibited by lithium ions in an *uncompetitive* manner.[30] This inhibition has been more closely investigated for IMPase due to its central role in inositol re-cycling described above, and may provide a rationale for the valuable effects of lithium treatment in manic-depressive illness. Uncompetitive inhibition is very rare in metabolic regulation, and is defined as inhibition which results in parallel Lineweaver-Burk plots, as shown in *Fig 2-9*. In

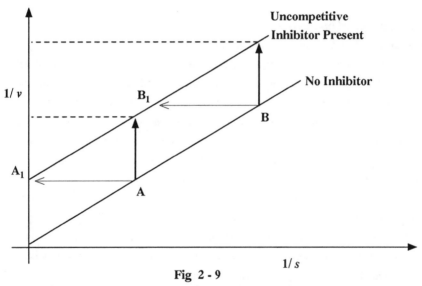

Fig 2 - 9

terms of mechanism this results from an inhibitor which does not bind to either the free enzyme, or to any form of the enzyme upstream of substrate binding which is a precursor of the substrate-bound form. The simplest mechanism which gives these results is therefore one in which the inhibitor binds only to the enzyme-substrate complex, and not to the free enzyme, although the mechanism of IMPase and its inhibition by lithium is probably more complex than this.[31] The unique consequences of this type of inhibition are illustrated in *Fig 2-9*. Starting from a substrate concentration of either A or B, addition of a fixed quantity of inhibitor causes a decrease in the velocity of the reaction, thick lines (*ie* an increase in 1/*v*) which is proportionally much greater at starting concentration A than at B, new velocity indicated by dotted lines. Thus it can be seen that uncompetitive inhibitors work best at low values of $1/s$ *ie* high substrate concentrations, and if the substrate concentration is increased, so is the degree of inhibition, exactly the inverse of the situation which occurs for competitive inhibitors. In addition, the increase in substrate concentration needed to return to the original velocity in the presence of the inhibitor is indicated by the thin arrows, resulting in new substrate concentrations of A_1 and B_1. It can be seen that for high starting substrate concentrations *eg* A, the rise needed to overcome the inhibition becomes infinite, *ie* $1/s = 0$, and restoration of the original velocity therefore becomes impossible. The consequences of this type of inhibition in a metabolic pathway were first pointed out by Cornish-Bowden,[32] and in essence the result is that the faster the pathway is turning over, the more effective will be its inhibition by an uncompetitive inhibitor. This uncompetitive effect of lithium on the inositol recycling pathway, coupled with the fact that the brain does not have free access to plasma inositol may explain the observed effects of lithium in manic-depressive illness.

2.6 Lithium and Manic-Depression

In the form of its citrate and carbonate salts lithium has been used successfully for over 30 years for the treatment of affective disorders, particularly manic-depressive illness (bipolar depression).[33] The clinical use of lithium initially focused on its ability to ameliorate mania, but it is now clear that after a lag period of one to two weeks lithium treatment also reduces the severity, and incidence, of both the manic and depressive phases of bipolar depression. The major limitation of lithium as a drug is its very low therapeutic ratio, or therapeutic window. To achieve the desired therapeutic effect a plasma concentration of 0.5 to 1.2 mM lithium is required, but serious side effects occur at plasma concentrations of 2.0 mM and above, and plasma levels of over 3.0 mM result in severe toxic effects including coma and eventually death. This profile makes the drug difficult to use in safety, and requires regular monitoring of plasma lithium levels on initiation of treatment. Safer and easier to use anti-manic drugs could result from elucidation of the mechanism of action of lithium, which has remained elusive despite numerous proposals.[34]

Following the discovery that lithium acts as an uncompetitive inhibitor of IMPase *in vivo*,[35] Berridge *et al* suggested that lithium inhibition of IMPase could reduce the supply of inositol available for PIP$_2$ synthesis in stimulated cells, and thus tone down receptor signalling *via* the 1,4,5-IP$_3$ second messenger pathway.[36] This idea, coupled with the uncompetitive nature of the inhibition of IMPase and the restricted access of cells in the brain to plasma inositol could provide an overall explanation of the observed anti-manic effects of lithium.[30]

Cells in the periphery have relatively free access to plasma inositol, so blockade of the inositol recycling system by inhibition of IMPase would be expected to have little effect on

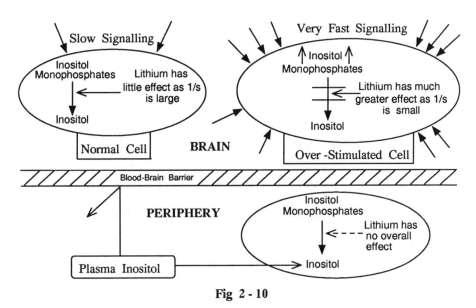

Fig 2 - 10

the availability of inositol, and on PIP_2 levels available for receptor controlled hydrolysis. In the brain however, blockade of IMPase would be expected to reduce the overall availability of inositol, and thus PIP_2, inside the cell as the blood-brain barrier precludes access of brain cells to plasma inositol. Indeed an accumulation of inositol monophosphates and a depletion of free inositol levels can be demonstrated in the brains of lithium-treated rats. This would explain why the observed anti-manic and toxic effects of lithium are CNS related, despite the fact that the $1,4,5$-IP_3 second messenger system exists throughout the body, and thus lithium treatment could be expected to inhibit cell-cell signalling in all tissues.

The inhibition of IMPase by lithium is uncompetitive, and thus as the flux through the inositol recycling system increases, with a concomitant increase in the levels of the inositol monophosphates, the more effectively lithium inhibits IMPase (*ie* as $1/s$ becomes smaller the relative change in $1/v$ becomes larger, *Fig 2-9*). Put another way, lithium should inhibit IPMase most effectively in cells where the receptors controlling the production of $1,4,5$-IP_3 are working hardest. This can be interpreted to mean that cells which are signalling normally will be little affected by lithium, as inositol monophosphate levels are low. Over-stimulated cells however which are signalling very rapidly, will have their signalling systems progressively inhibited by lithium with the degree of inhibition increasing as the stimulation increases, *Fig 2-10*. If we now assume that bipolar depression is due to the over-activity of one population of neurones and their associated signalling system during the manic phase, and to the over-activity of a second population of neurones during the depressive phase, the effect of lithium in improving the symptoms of both aspects of manic-depressive illness can be explained.

The above rationalisation that inhibition of IMPase is responsible for the observed effects of lithium in manic-depressive illness although attractive, has not been proven at this time. The development of new anti-manic drugs based on this working hypothesis of IMPase inhibition has however been undertaken by several pharmaceutical companies, despite uncertainties over whether a "normal" competitive inhibitor could exhibit the same activity profile as the uncompetitive inhibitor lithium. The results of these studies disclosed to date, see Chapter 8, indicate that the major problem to be solved in this area is the lack of penetration of the synthetic inhibitors into the brain, and until potent selective inhibitors of IMPase which are able to penetrate the brain are available, the above hypothesis must remain speculative.

References

1)   B.Alberts, D.Bray, J.Lewis, M.Raff, K.Roberts and J.D.Watson, "Molecular Biology Of The Cell", Chapter 12, 1989, Garland, London.

2)   R.M.Evans, *Science*, 1988, **240**, 889; U.Gehring, *Trends Biochem.Sci.*, 1987, **12**, 399.

3)   S.H.Snyder, *Scientific American*, 1985, **253**, 132; J.Findlay, *Chemistry in Britain*, 1991, 724.

4)   M.S.Goligorski, *Contemp.Issues Nephrol.*, 1991, **23**, 1.

5)   B.L.Blazer-Yost, S.Goldfarb and F.N.Ziyadeh, *Contemp.Issues Nephrol.*, 1991, **23**, 339; R.C.Harris, *ibid*, 1991, **23**, 365.

6)   R.L.Rawls, *Chem.Eng.News*, 1987, **65**, 26.

7)   A.M.Spiegel, *Medicinal Res.Rev.*, 1992, **12**, 55.

8)   C.W.Taylor, *Biochem.J.*, 1990, **272**, 1.

9)   H.R.Bourne, D.A.Sanders and F.McCormic, *Nature*, 1991, **349**, 117.

10)   H.E.Ives, *Cellular Signalling*, 1991, **3**, 491.

11)   A.G.Gilman, *Annual Rev.Biochem.*, 1987, **56**, 615; L.Stryer and H.R.Bourne, *Annual Rev.Cell Biol.*, 1986, **2**, 391.

12)   S.J.Hill and D.A.Kendall, *Drug News and Perspectives*, 1992, **5**, 39; A.A.Abdel-Latif, *Cellular Signalling*, 1991, **3**, 371.

13)   M.Schramm and Z.Selinger, *Science*, 1984, **225**, 1350; A.Levitzki, *Science*, 1988, **241**, 800.

14)   M.J.Berridge, *Proc.R.Soc.Lond.(Biol.)*, 1988, **234**, 359; " Transmembrane Signalling, Intracellular Messengers and Implications for Drug Development ", 1991, S.R.Nahorski ed., Wiley, New York.

15)   M.R.Hokin and L.E Hokin, *J.Biol.Chem.*, 1953, **203**, 967; L.E.Hokin, *Ann.Rev. Biochem.*, 1985, **54**, 205.

16) R.H.Michell, *Biochim.Biophys.Acta.*, 1975, **415**, 81.

17) Y.Nishizuka, *Nature*, 1984, **308**, 693; M.J.Berridge and R.F.Irvine, *Nature*, 1984, **312**, 315; C.P.Downes, *Trends Neurosci.*, 1986, **9**, 394; S.R.Nahorski, *Trends Neurosci.*, 1988, **11**, 444; N.N.Osbourne, A.B.Toblin and H.Ghazi, *Neurochem.Res.*, 1988, **13**, 177; R.F.Irvine, R.M.Moor, W.K.Pollock, P.M.Smith and K.A.Wreggett, *Phil.Trans.R.Soc.Lond.(Biol.)*, 1988, **320**, 281;

18) M.J.Berridge, *Annual.Rev.Biochem.*, 1987, **56**, 159.

19) S.Supattapone, P.F.Worley, J.M.Baraban and S.H.Snyder, *J.Biol.Chem.*, 1988, **263**, 1530.

20) T.Furuichi, S.Yoshikawa, A.Miyawaki, A.Wada, N.Maeda and K.Mikoshiba, *Nature*, 1989, **342**, 32.

21) G.A.Mignery, C.L.Newton, B.T.Archer III and T.C.Sudhof, *J.Biol.Chem.*, 1990, **265**, 12679.

22) C.D.Ferris, R.L.Huganir, S.Supattapone and S.H.Snyder, *Nature*, 1989, **342**, 87.

23) R.D.Burgoyne and T.R.Cheek, *Trends Biochem.Sci.*, 1991, **16**, 319.

24) S.B.Shears, *Pharmac.Ther.*, 1991, **49**, 79.

25) C.Erneux and K.Takazawa, *Trends Pharm.Sci.*, 1991, **12**, 174.

26) R.H.Michell, *Nature*, 1986, **324**, 613; S.R.Nahorski and I.Batty, *Trends Pharm.Sci.*, 1986, **7**, 83; M.D.Houslay, *Trends Biochem.Sci.*, 1987, **12**, 133; A.Luckhoff and D.E.Clapham, *Nature*, 1992, **355**, 356.

27) E.Neher, *Nature*, 1992, **355**, 298.

28) R.F.Irvine, *BioEssays*, 1991, **13**, 419.

29) L.M.Lewin, Y.Yanna, S.Sulimovici and P.F.Kraicer, *Biochem.J.*, 1976, **156**, 375.

30) A.H.Drummond, *Trends Pharm.Sci.*, 1987, **8**, 129; S.R.Nahorski, C.I.Ragan and R.A.J.Challiss, *Trends Pharm.Sci.*, 1991, **12**, 297.

31) J.K.Shute, R.Baker, D.C.Billington and D.Gani, *J.Chem.Soc.Chem.Commun.*, 1988, 626; G.R.Baker and D.Gani, *Bioorg.Med.Chem.Lett.*, 1991, **1**, 193.

32) A.Cornish-Bowden, *FEBS Lett.*, 1985, **203**, 3.

33) "Lithium Research and Therapy", F.N.Johnson ed, 1975, Academic Press.

34) N.H.Hendler in "Handbook of Psychopharmacology", L.L.Iversen, S.D.Iversen and S.H.Snyder eds, 1978, **14**, 233.

35) L.M.Halcher and W.R.Sherman, *J.Biol.Chem.*, 1980, **255**, 10896; W.R.Sherman, B.G.Gish, M.P.Honchar and L.Y.Munsell, *Fed.Proc.*, 1986, **45**, 2639.

36) M.J.Berridge, P.F.Downes and M.R.Hanley, *Biochem.J.*, 1982, **206**, 587.

Chapter 3

General Synthetic Considerations

3.1 Introduction

This chapter aims to introduce the basic problems posed by the synthesis of the inositol phosphates, and the methods which have been developed and exploited to solve these problems.

The synthesis of a given inositol phosphate, and of many of the analogues described in this book, presents three main synthetic problems, and thus the overall synthetic approach can be divided into three related operations: (1) the preparation of a suitably protected inositol derivative, having free hydroxyl groups at the desired positions; (2) phosphorylation in an efficient manner with a reagent bearing suitable phosphate-protecting groups (a major problem for *vic* diols in particular due to steric hindrance of the second phosphorylation, and the ready formation of 5-membered cyclic phosphates); (3) removal of the protecting groups in an efficient manner without migration of phosphate groups to adjacent free hydroxyl groups, or the formation of side products which may necessitate purification of the final product by HPLC methods. For the synthesis of optically active inositol phosphates and analogues, resolution of a suitable intermediate by efficient techniques during phase (1), or the use of chiral starting materials becomes important, *Fig 3-1*.

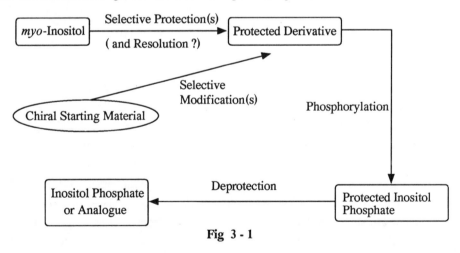

Fig 3 - 1

3.2 Synthesis of Protected Inositol Derivatives

3.2.1 myo-Inositol as a Starting Material

Due to the ready commercial availability of pure *myo*-inositol, most syntheses use the parent cyclitol as a starting material. The inherent symmetry of the inositol system (see Chapter 1) is an added benefit in the design of many synthetic routes from this material.

Reaction of inositol with cyclohexanone,[1] or more efficiently with a cyclohexanone precursor under acidic conditions leads to the formation of the three isomeric bisacetals shown, (1), (2) and (3) *Fig 3-2*, together with the trisacetal.[2,3] Each of the three isomeric bisacetals may be obtained in pure form by a combination of crystallisation and chromatography (acetal (1) can be crystallised directly from the reaction work-up, and (2) and (3) are recovered by chromatography). All of the bisacetals, and the trisacetal, give the monoacetal (4) on mild acid hydrolysis due to selective cleavage of the less stable *trans* acetal group.

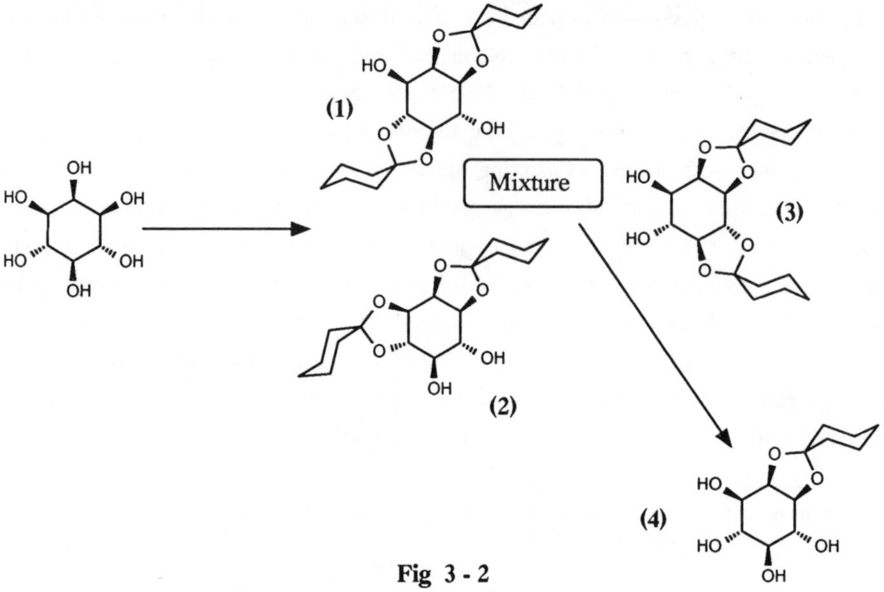

Fig 3 - 2

Each of the bisacetals (1) - (3) has proved valuable for the synthesis of inositol phosphates, due to the relative stability of the acetal protecting group,[4] and the differential reactivities displayed by their free hydroxyl groups.[5] Detailed NMR analysis coupled with molecular modelling studies indicate that each of the bisacetals is relatively conformationally rigid, but none exist in the classic chair conformation. This is presumably due to the constraints imposed by the fused 5-membered acetal rings on the cyclohexane system. Under suitably controlled conditions each of the bisacetals may be selectively alkylated or acylated, and some examples are shown in *Fig 3-3*. These selective protections, coupled with selective hydrolysis of the less stable *trans* acetal group, and the possibility of effecting further selective protection of the resulting *vic* diol has led to synthetic sequences of the type shown in *Fig 3-4* being successfully employed.

In the synthesis of racemic 1,4,5-inositol trisphosphate (1,4,5-IP$_3$) shown in *Fig 3-4*, Frazer-Reid and Yu [6] achieved 97% chemoselectivity in the conversion of the bisacetal (3)

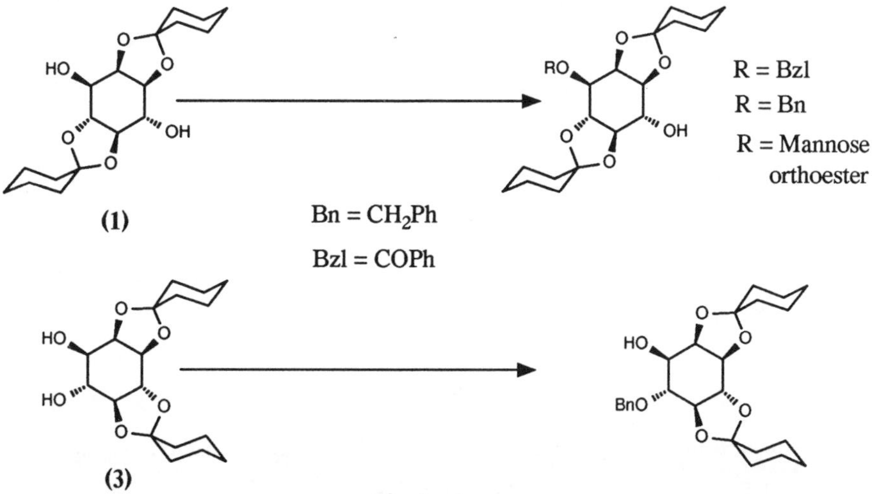

R = Bzl
R = Bn
R = Mannose
orthoester

(1)   Bn = CH₂Ph

Bzl = COPh

(3)

Fig 3 - 3

into (5) *via* the use of a tin-mediated benzylation.[7] Subsequent selective hydrolysis of the *trans* acetal gave the racemic triol (6). Phosphorylation of this triol using their two-step, one-pot protocol involving N,N-diisopropyl dibenzyl phosphoramidite, followed by oxidation with mCPBA, gave the fully protected racemic compound (7), in excellent yield. Although deprotection of (7) was not reported by these authors, presumably hydrogenolysis followed by acid treatment would give *rac*-1,4,5-IP₃, in analogy to the deprotection of the optically active material which had been previously reported.

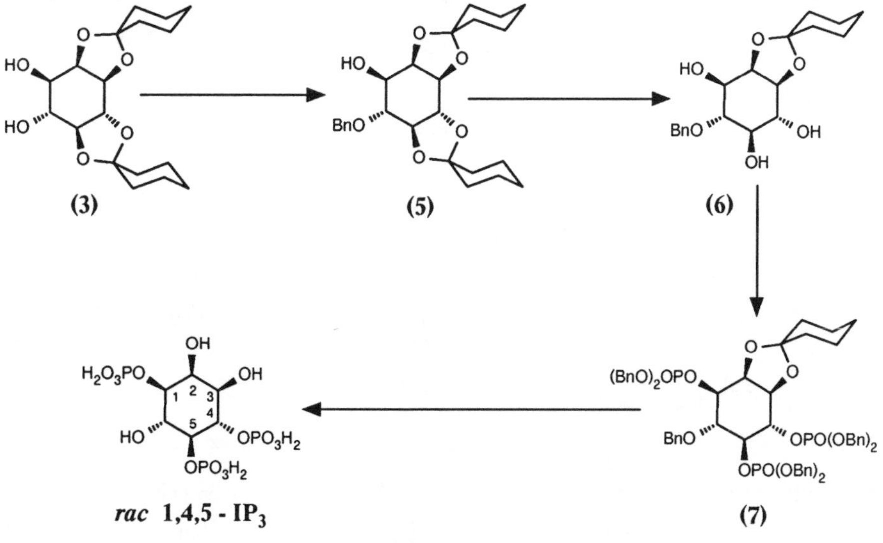

(3)   **(5)**   **(6)**

rac 1,4,5 - IP₃   **(7)**

Fig 3 - 4

An example of the selective protection of a single hydroxyl group in a diol (8) resulting from the hydrolysis of a *trans* acetal is given in *Fig 3-5*, producing the alcohol (9), which on allylation and subsequent hydrolysis of the remaining *cis* acetal protecting group liberates the protected 1,2-diol (10).

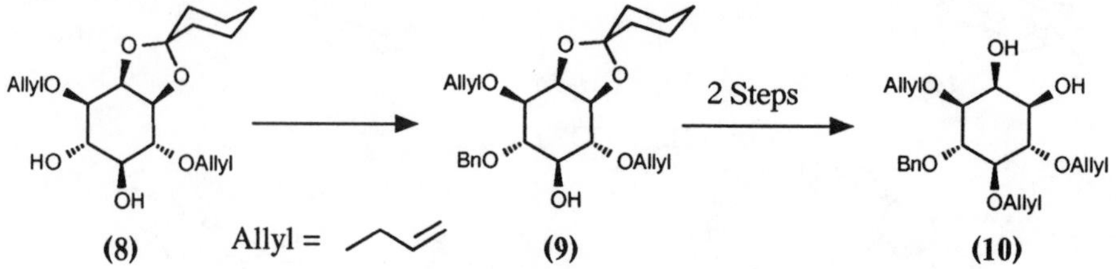

Fig 3 - 5

A general feature of this type of inositol 1,2-diol is the higher reactivity of the equatorial hydroxyl group at the 1- position compared with the axial hydroxyl group at the 2-position. Many authors have exploited this to provide selectively protected inositols having either the 1- or the 2- hydroxyl group free. As most of the naturally occurring inositol phosphates have a free hydroxyl group at the 2-position, and a phosphate group at the 1-position, the selective incorporation of a temporary protecting group at the more reactive 1-position, followed by protection with a more robust group at the 2-position, and removal of the temporary group has been widely used. Some examples of this type of manipulation are given in *Fig 3-6*.

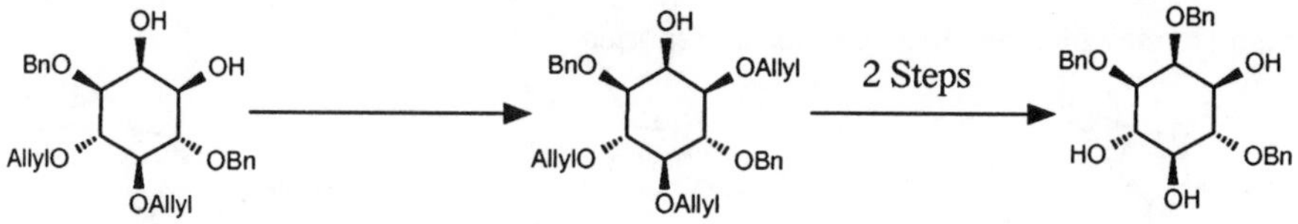

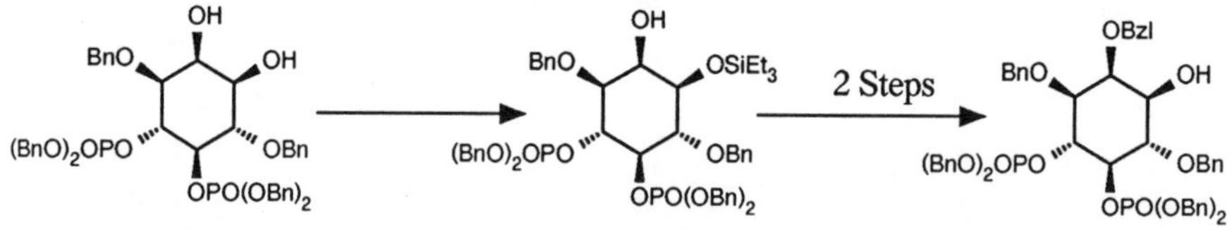

Fig 3 - 6

Selective allylation at the 1-position of the inositol 1,2-diol (11) gives (12), one of the first examples of this type of selectivity to be reported,[2] and subsequent benzylation of the free hydroxyl group in (12) followed by removal of the three allyl groups gives the alcohol (13), a key intermediate in a synthesis of 1,4,5-IP$_3$.[8] A more recent example of this approach using alternative protecting groups involves the selective silylation of the 1,2-diol (14), giving (15), which on benzoylation and removal of the silyl group gives the alcohol (16).[9]

In addition to the cyclohexanone acetals described above, other acetals including the isopropylidine and cyclopentylidine acetals have been studied and used in synthesis. The overall reactivity patterns of these acetals is very similar to the cyclohexylidine compounds.[8,10,11]

Treatment of *myo*-inositol with triethylorthoformate at high temperatures, in the presence of an acid catalyst results in the formation of the crystalline orthoformate derivative (17), *Fig 3-7*.[12] This derivative provides an interesting protected inositol, in which the 1, 3, and 5 hydroxyl groups are protected simultaneously, and in addition the normal axial/equatorial relationship of the remaining free hydroxyl groups has been reversed. The spatial

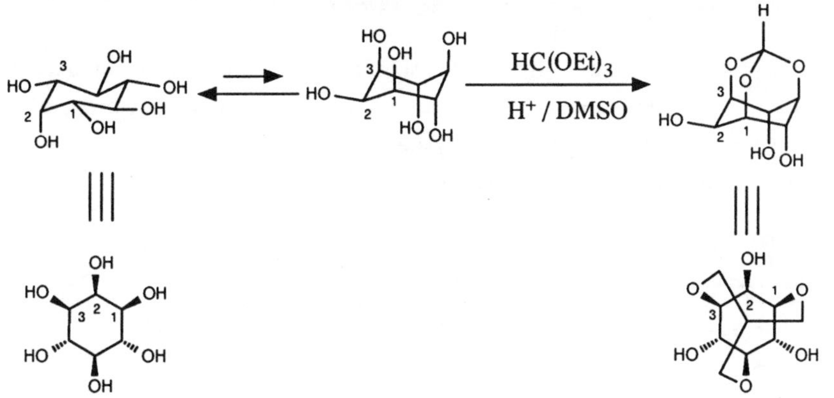

myo - Inositol

(17)   Inositol
Orthoformate

Fig 3 - 7

juxtaposition of the axial hydroxyl groups in (17) allows highly selective monoalkylations and monophosphorylations to be performed, *Fig 3-8*.[13] The very high regioselectivity of this reaction, coupled with the high degree of monoalkylation observed are presumably due to

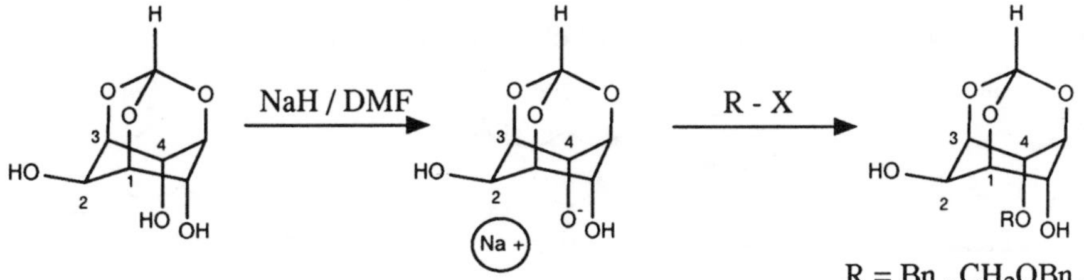

(17)

R = Bn , CH$_2$OBn ,
Allyl , *etc*

Fig 3 - 8

internal coordination in the intermediate anion, *ie* the existence of a chelated intermediate of the type shown in *Fig 3-8*. The observation that changes in either solvent or counter-ion lead to loss of selectivity supports this type of intermediate being involved.[14] An example of the use of this intermediate is shown in *Fig 3-9*, where selective allylation of (17) gives a high yield of the monoallyl derivative (18). Benzylation of the two remaining free hydroxyl groups in (18) then gives (19), which on removal of the allyl and orthoformate groups gives the tetrol (20), a key intermediate for the synthesis of 1,3,4,5-inositol tetrakisphosphate.[14]

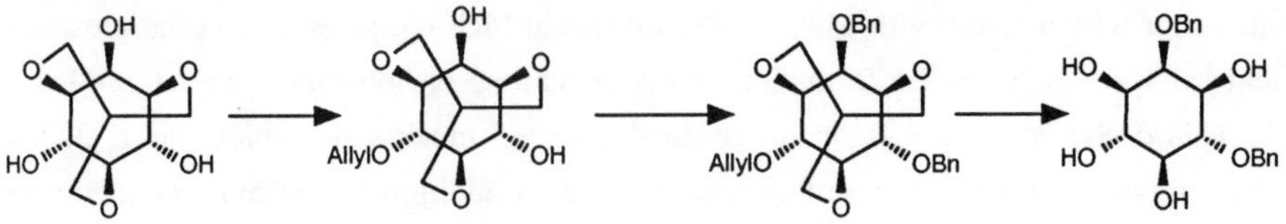

Fig 3 - 9

A selective protection of the 3,4 hydroxyl groups of acetal (4) is possible, by treatment with 1,3-dichloro-1,1,3,3- tetraisopropyldisiloxane (TIPS) resulting in selective formation of the diol (21), which may be selectively benzoylated to give the alcohol (22), an intermediate in the synthesis of racemic inositol 4-phosphate, *Fig 3-10*.[15]

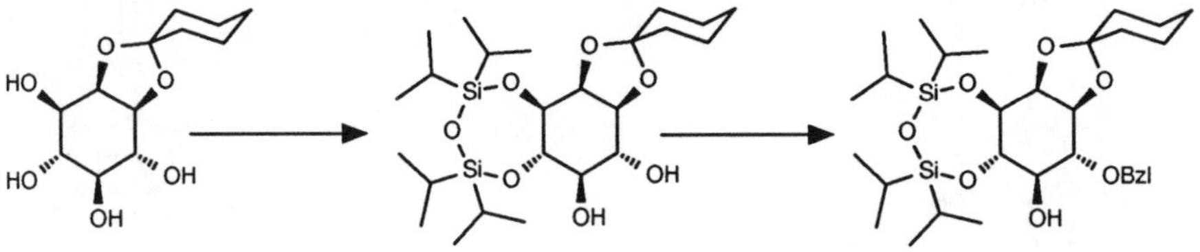

Fig 3 - 10

Some selectivity can be obtained by careful control of the reaction conditions in the direct benzoylation of *myo*-inositol, which can provide for example the 1,3,4,5-tetrabenzoate (23), *Fig 3-11*, in 33% isolated yield.[16] Benzylation of the two free hydroxyl groups of (23) using benzyl trichloroacetimidate in the presence of trifluoromethanesulphonic acid then provides the fully protected intermediate (24). Hydrolysis of the four ester groups in (24) then

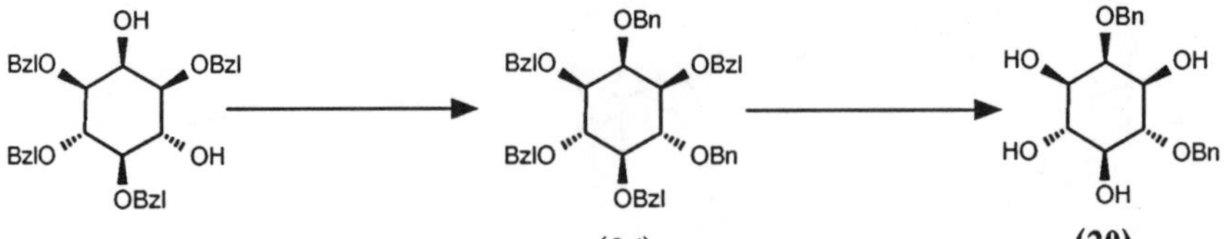

Fig 3 - 11

produces the racemic tetrol (20), *cf Fig 3-9*, used in the synthesis of 1,3,4,5-IP$_4$.

A number of other interesting selective reactions based on *myo*-inositol have been reported, which have not been exploited in the syntheses of inositol phosphates as yet, and some recent examples are given below. These examples have been selected to illustrate the isolation of protected derivatives having substitution patterns which are inaccessible or tedious to achieve by the more conventional strategies given above, and do not represent an exhaustive coverage of the literature.

Reaction of *myo*-inositol with 2.5 equiv. of TIPS chloride in pyridine at 25 degrees C gives a 66% yield of the symmetrical derivative (25), with no other observable inositol derivatives, *Fig 3-12*[15].

Fig 3 - 12   **(25)**

Treatment of *myo*-inositol with triethylborane gives the hexa-O-diethylboryl derivative (26), which is hexane soluble.[17] Reaction of (26) with 1 equiv. of the tri-*n*-butylstannyl enolate of pentane-2,4-dione, results in the formation of a partially stannylated derivative, which may be selectively alkylated at the 1-position. Thus treatment of the stannane with benzyl bromide gives a 63% yield of 1-O-benzyl inositol, and with allyl bromide gives 53% yield of the 1-O-allyl derivative, *Fig 3-13*. By altering the degree of stannylation of the intermediate

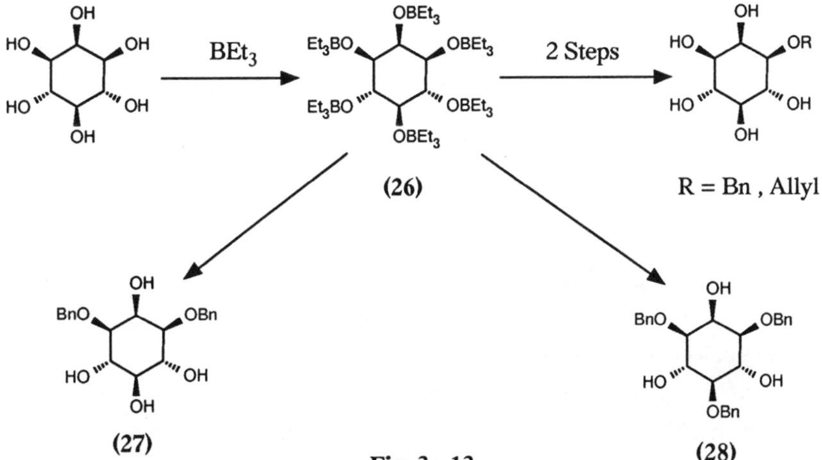

(27)   **Fig 3 - 13**   **(28)**

inositol derivative, and the quantity of alkylating agent used in the final step, 1,3-di-O-benzyl inositol (27) (48% yield) or 1,3,5-tri-O-benzyl inositol (28) (27% yield) can be obtained as the major reaction product.

The trisbenzyl orthoformate (29), obtained by benzylation of (17) *Fig 3-7*, may be selectively cleaved to give either the acetal (30) or the acetal (31) by treatment with diisobutylaluminium hydride or triethylaluminium, respectively, *Fig 3-14*.[18] The yield for each of these conversions is *ca* 90%.

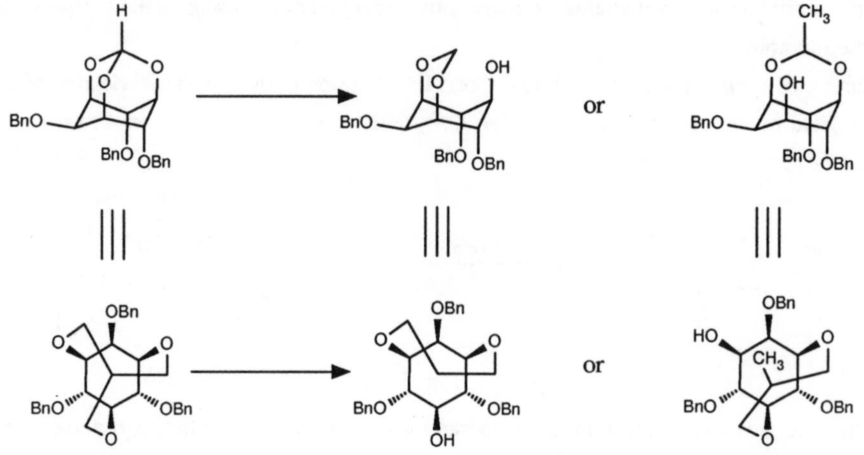

(29)   (30)   (31)

Fig 3 - 14

3.2.2 Other Starting Materials

This section will cover non-chiral starting materials, the use of chiral starting materials is covered in Section 3.2.4.

Two fundamental approaches to the *myo*-inositol skeleton based on the use of cyclohexadiene diols have been developed, one from benzene *via* the *cis*-diol (32), and one from the corresponding *trans*-diol.

Treatment of the *meso*-diol (32), obtained from microbial oxidation of benzene, with phosgene produces the cyclic carbonate (33), *Fig 3-15*.[19] Epoxidation of the diene (33) with mCPBA gives the *alpha*-epoxide (34) as the major product (>4:1), which undergoes regioselective ring opening on treatment with benzyl alcohol under acidic conditions to give the alcohol (35). Benzylation of the free hydroxyl group then gives the fully protected derivative. Stereoselective epoxidation of the remaining double bond is achieved by removal of the carbonate group to expose the *cis* diol function, followed by hydroxyl-directed epoxidation using mCPBA at pH 8.8 to produce the *beta*-epoxide, and reprotection of the diol function as the acetonide to give (36). Less than 10% of the unwanted *alpha*-epoxide is formed in this sequence. As predicted by conformational analysis, regioselective epoxide opening of (36) with Ley's sterically unencumbered hydroxide equivalent DDEO gives the

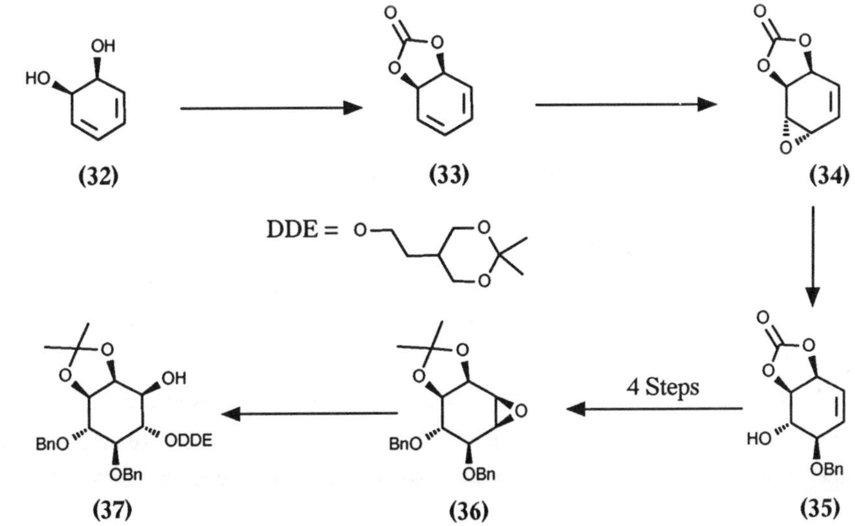

Fig 3 - 15

desired alcohol (37), which has all of the relative stereochemistry of *myo*-inositol established, in acceptable yield. Intermediate (37) can be used for the synthesis of racemic 1,4,5-IP$_3$,[19,20] and opening of the epoxide (36) with other nucleophiles gives access to a range of 1,4,5-IP$_3$ analogues.[20]

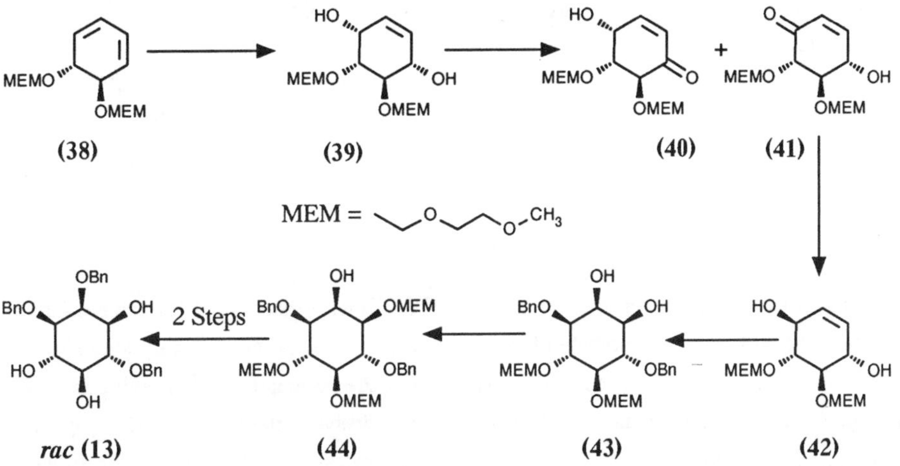

Fig 3 - 16

The other general approach to the *myo*-inositol skeleton involves [4 + 2] addition of singlet oxygen to the protected *trans* diol (38), which produces the protected tetrol (39), *Fig 3-16*.[21] Oxidation of (39) to a separable mixture of (40) and (41), followed by reduction of the desired isomer (41) with NaBH$_4$/CeCl$_3$ then gives the diol (42). Benzylation of (42) followed by *cis* hydroxylation using the catalytic OsO$_4$ system then produces the protected *myo*-inositol derivative (43). Selective protection of the equatorial position in (43) gives (44) as the major product, which on benzylation and removal of the MEM groups gives the known racemic triol (13), an intermediate in the synthesis of 1,4,5-IP$_3$.

3.2.3 Resolution of Protected Inositols

The classical resolution method of formation and separation of diastereomeric salts has found little success in the inositol area. During a synthesis of the enantiomers of inositol 1-phosphate for example, resolution of (45) was possible, by conversion to the acid oxalate (46) followed by salt formation with chiral bases leading to the resolved pentaacetates (47) and (48), *Fig 3-17*. The problem is however that a different chiral base is required for each enantiomer, and multiple recrystallisations are needed to obtain optically pure material.[22,23]

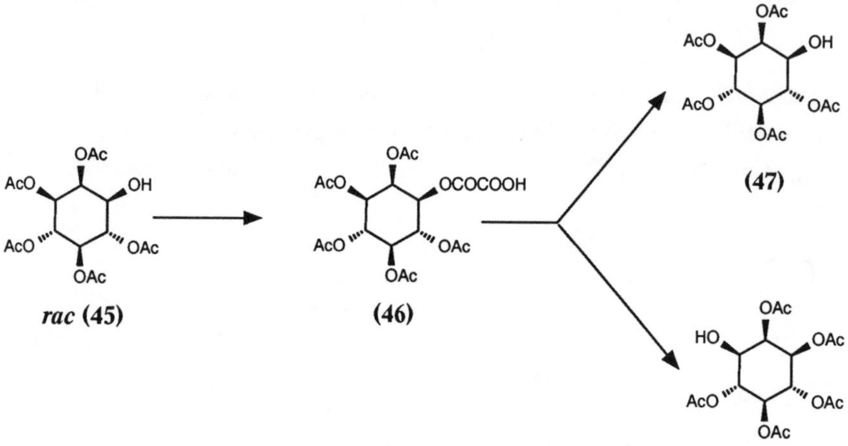

Fig 3 - 17

Limited success has been reported in the use of enzymes (esterases) for the resolution of protected inositols. Thus cholesterol esterase shows stereo- and regiospecificity in the hydrolysis of the racemic diacetate (49) derived from the bisacetal (3),[24] allowing isolation of the optically active monoacetate (-)-(50), and after hydrolysis, the diol (+)-(51), *Fig 3-18*, an intermediate in the synthesis of (-)-1,4,5-IP$_3$. The *meso* diester (52), prepared in three steps from the orthoformate (17), is selectively hydrolysed by pig liver esterase to give the diol (53) in 95% enantiomeric excess, *Fig 3-19*.[25] This diol played a key role in a synthesis of optically active 1,3,4,5-IP$_4$.

The most widely used and successful resolution method involves the conversion of a partially protected racemic (or *meso*) inositol derivative, having a free hydroxyl group, into a pair of

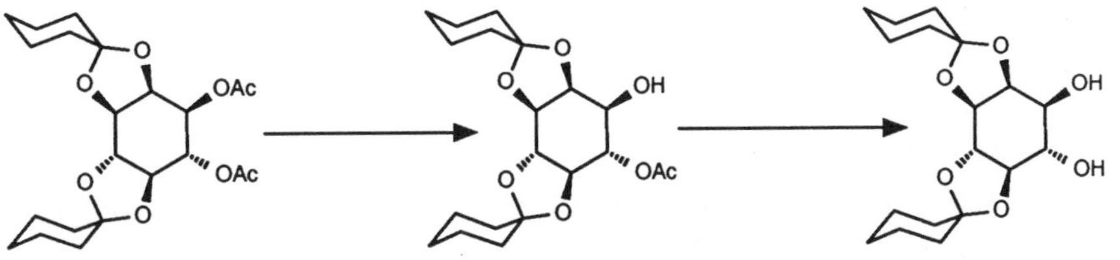

rac (49) **(-)-(50)** **(+)-(51)**

Fig 3 - 18

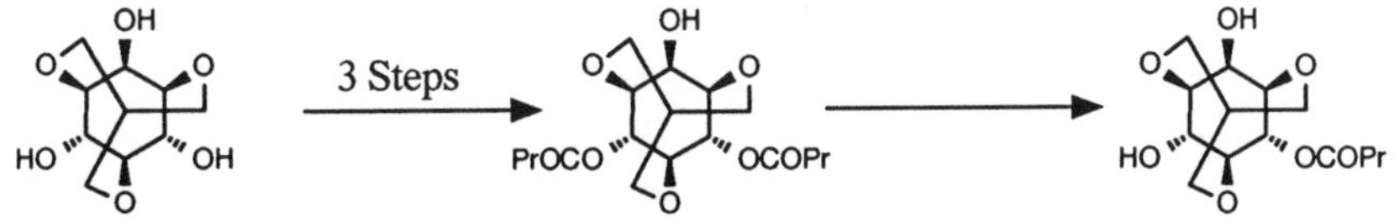

(17) **(52)** **(-)-(53)**

Fig 3 - 19

diastereomeric esters *via* esterification with a chiral acid or acid derivative. These diastereomeric esters may then be separated by normal chromatographic methods, or by crystallisation in favourable cases. Hydrolysis of the esters then gives the enantiomers of the starting racemic protected alcohol, *Fig 3-20*, and the chiral acid is recovered. This general approach is particularly valuable as the *diastereomeric purity* of the separated esters can be accurately determined by HPLC, which reflects the *optical purity* of the final alcohols

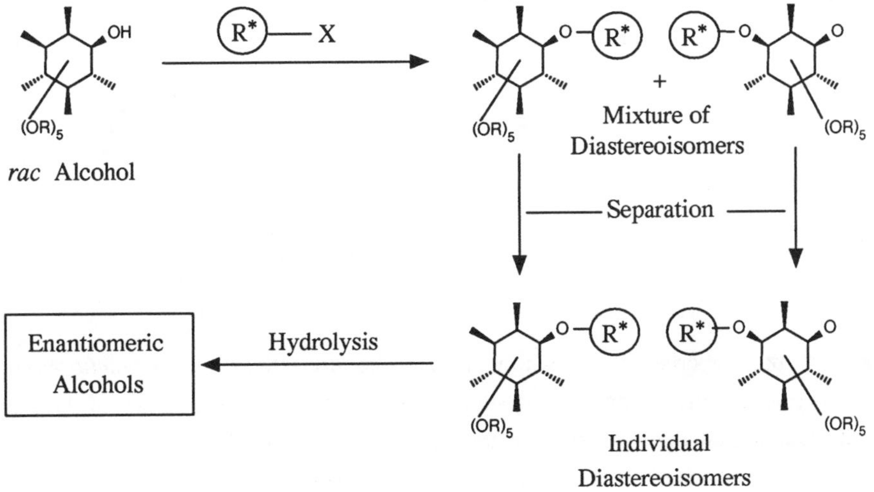

Fig 3 - 20

obtained, and as X-ray crystallography of the ester derivatives may be used to determine unambiguously the absolute configuration of the individual enantiomeric alcohols derived from them.

Four variations on the above general approach have been reported, using camphanate esters (a); menthoxyacetate esters (b); orthoesters of D-mannose (c); and tartrate esters (d), *Fig 3-21.*[26] An additional advantage of this approach is that in some cases the formation of a

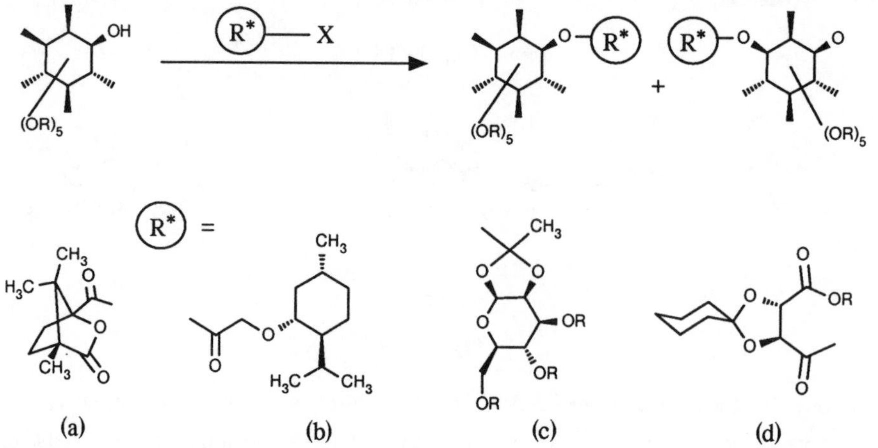

Fig 3 - 21

single ester is possible from derivatives having more than one free hydroxyl group, thus potentially reducing the number of protecting group manipulations necessary during a given synthesis. An example of this is given in *Fig 3-22*, where diol (54) reacts selectively at the less hindered equatorial hydroxyl group giving a pair of separable diastereomeric mono esters as shown.[27] Taking this approach one step further, tartrate derivatives such as D-(55) and

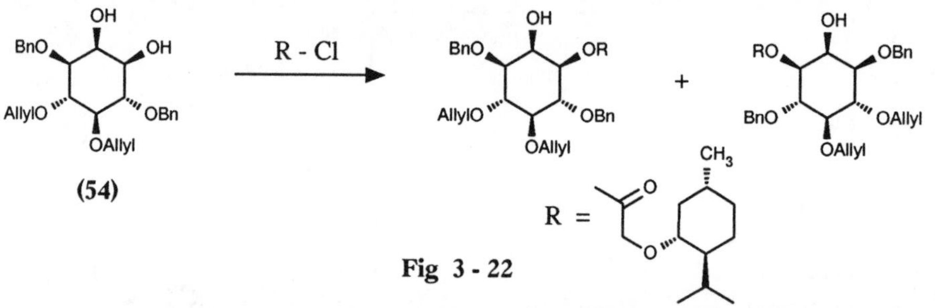

Fig 3 - 22

L-(56) have been reported to react with various inositol derivatives *eg* (57) *enantioselectively* as well as regioselectively, that is they show a preference for one enantiomer of the starting alcohol used, and thus give mixtures of the diastereomeric esters (58) and (59) in a non 1:1 ratio, *Fig 3-23.*[28] In the best cases up to 98:2 ratios were obtained in 57% chemical yield, which represents a *direct* diastereomeric excess of 96% without any physical separation of

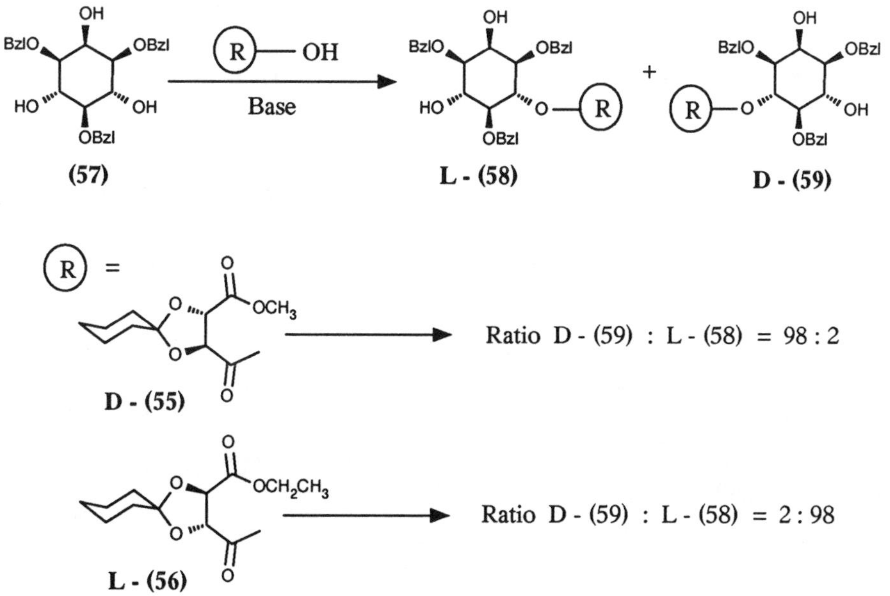

Fig 3 - 23

the diastereomers. By varying the structure of the tartrate and the enantiomer of the tartrate reagent used, each enantiomer of the inositol derivative could be obtained in excess. The diastereomeric ratios obtained on reaction with the tartrate derivative D-(55) with a series of inositol derivatives are given in *Fig 3-24*.

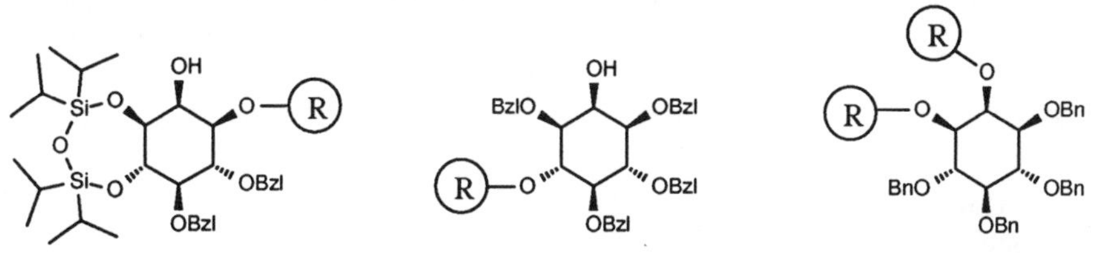

Ratio D : L = 83 : 17   Ratio D : L = 86 : 14   Ratio D : L = 84 : 16

Fig 3 - 24

Of the methods outlined above, the most generally applicable at this time is probably still the use of camphanate esters,[26] as: (1) (-) camphanic acid chloride is a stable crystalline reagent, readily available in high optical purity from many commercial sources; (2) the camphanate esters formed are usually readily separated on a large scale by conventional chromatography on silica gel (*eg* flash chromatography); (3) HPLC analysis of the separated esters gives the diastereomeric excess directly, and thus the enantiomeric excess of the alcohols obtained;[29] (4) the camphanates are often highly crystalline, and as the *absolute* configuration of the

(-)-camphanic acid is known, X-ray crystallographic structure determination allows the *absolute* configuration of the final enantiomers to be established unambiguously;[29,30] (5) the route has been successfully used for a wide range of inositol derivatives.[26,31,32]

3.2.4 Chiral Starting Materials

An alternative to resolution of intermediates is to begin the synthetic sequence with a naturally occurring chiral material. A number of synthetic routes have been developed using naturally occurring chiral inositol, or inositol-like, derivatives as starting materials.

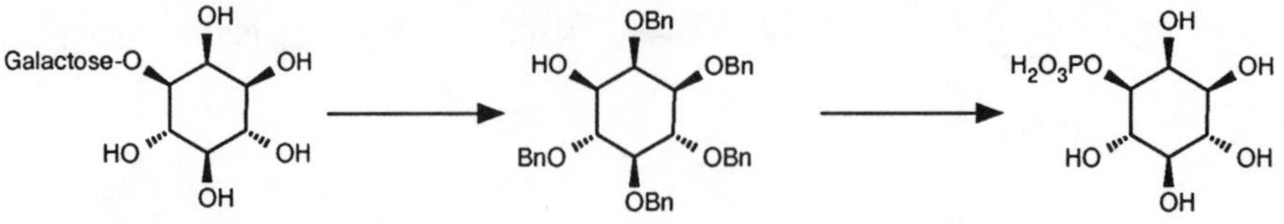

(60) Galactinol | **(61)** | **(62)** (-)-Inositol 1-phosphate

Fig 3 - 25

The first reported synthesis of (-)-inositol 1-phosphate used the naturally occurring galactinol (60) as a starting material, converting this to the pentabenzyl inositol (61) and then to (-)-inositol 1-phosphate (62), *Fig 3-25.*[33]
The naturally occurring methylated inositols L-quebrachitol (63) and D-pinitol (67) have both

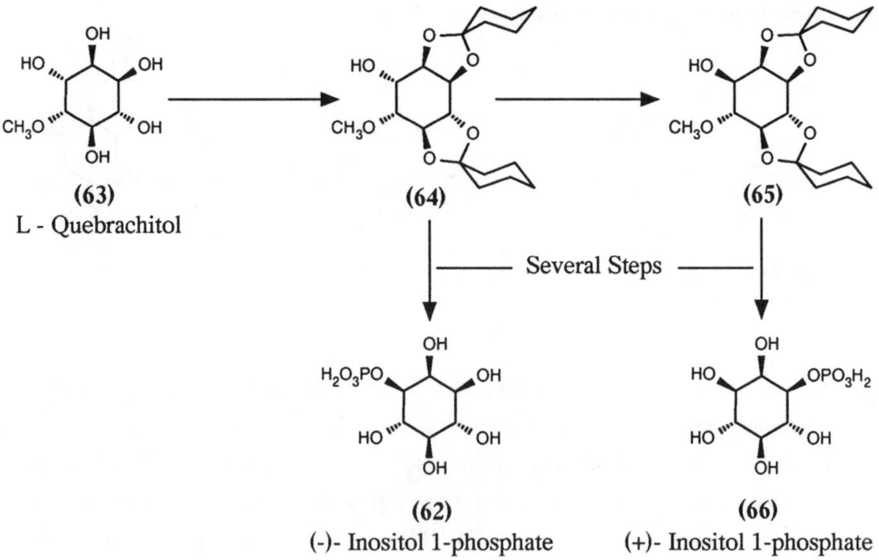

Fig 3 - 26

been used in the synthesis of inositol phosphates. L-Quebrachitol (63) may be converted into the more stable biscyclohexylidine acetal (one *cis* and one *trans* acetal group) (64), which has the correct stereochemistry for demethylation to lead directly to suitable intermediates for the synthesis of (-)-inositol 1-phosphate (62), *Fig 3-26*.[34] Alternatively, inversion of the free hydroxyl group in (64) leads *via* (65) to intermediates suitable for the synthesis of (+)-inositol 1-phosphate (66).[34]

L-Quebrachitol (63) and D-pinitol (67) have been converted into the individual enantiomers of 1,4,5-IP$_3$. The starting point in this strategy is the conversion of D-pinitol into D-*chiro*-inositol (68), and L-quebrachitol into L-*chiro*-inositol (69) by demethylation with HI, *Fig 3-27*.[35] D-*chiro*-inositol may then be converted into D-(-)-1,4,5-IP$_3$ (70), whilst repetition of the synthetic sequence using the L-*chiro* isomer gives L-(+)-1,4,5-IP$_3$, (71).

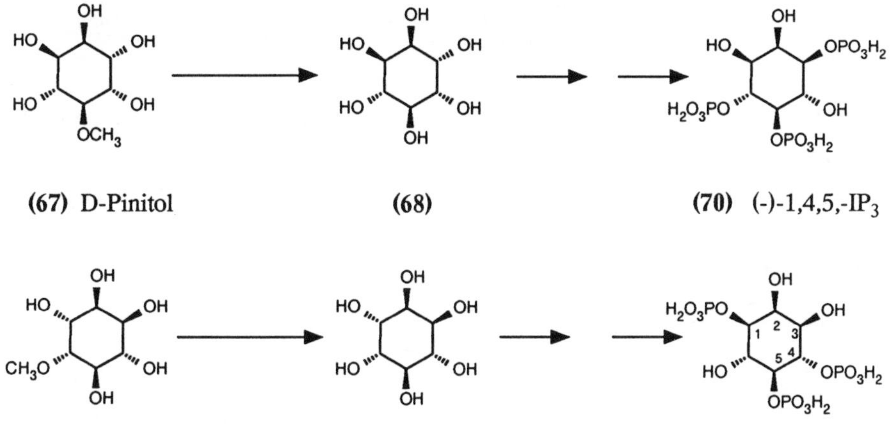

(67) D-Pinitol   **(68)**   **(70) (-)-1,4,5,-IP$_3$**

(63) L-Quebrachitol   **(69)**   **(71) (+)-1,4,5-IP$_3$**

Fig 3 - 27

A somewhat longer sequence of modifications is required to convert quinnic acid (72) into (-)-1,4,5-IP$_3$, *Fig 3-28*.[36] An 11-step sequence leads to the alcohol (73), which on removal of the two silyl groups gives the triol (74). Phosphorylation of (74) and deprotection then gives

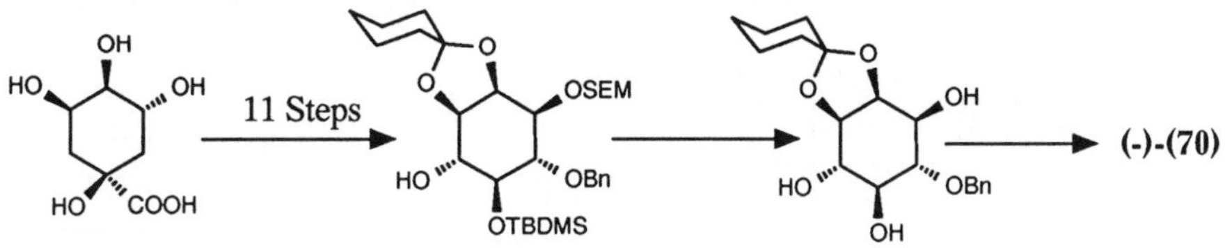

(72)   **(73)**   **(74)**

(-)-Quinnic acid

Fig 3 - 28

(-)-1,4,5-IP$_3$ (70).

3.3 Phosphorylation of Protected Inositols

The phosphorylation of inositol derivatives by conventional techniques using phosphorus (V) reagents such as diphenylchlorophosphate (DPCP) is severely limited for the following reasons : (1) the secondary hydroxyl groups of inositol are relatively unreactive to phosphorus (V) reagents like DPCP ; (2) the intermediate phosphate triesters formed have a

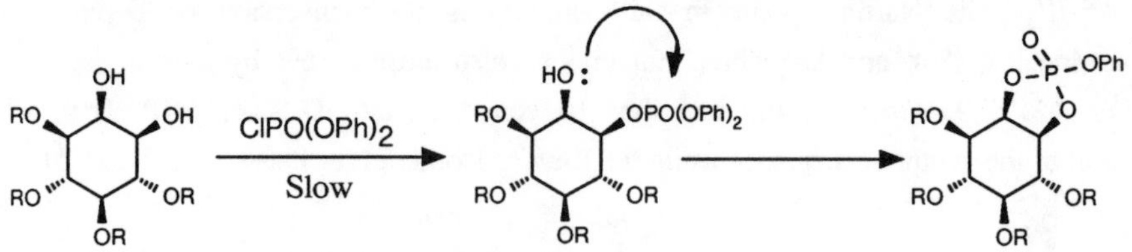

Fig 3 - 29

tendency to form 5-membered cyclic phosphates when neighbouring hydroxyl groups are available, rendering the bisphosphorylation of *vic* diols very difficult, and leading to mixtures of isomers, *Fig 3-29* ; (3) during deprotection, care must be taken to avoid migration of phosphate ester groups to adjacent free hydroxyl groups, again *via* 5-membered cyclic species, resulting in the formation of isomeric phosphates, *Fig 3-30*.

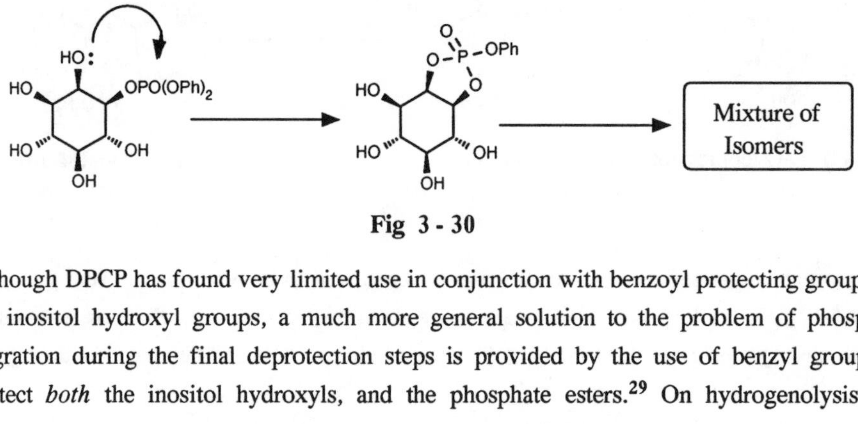

Fig 3 - 30

Although DPCP has found very limited use in conjunction with benzoyl protecting groups for the inositol hydroxyl groups, a much more general solution to the problem of phosphate migration during the final deprotection steps is provided by the use of benzyl groups to protect *both* the inositol hydroxyls, and the phosphate esters.[29] On hydrogenolysis, the

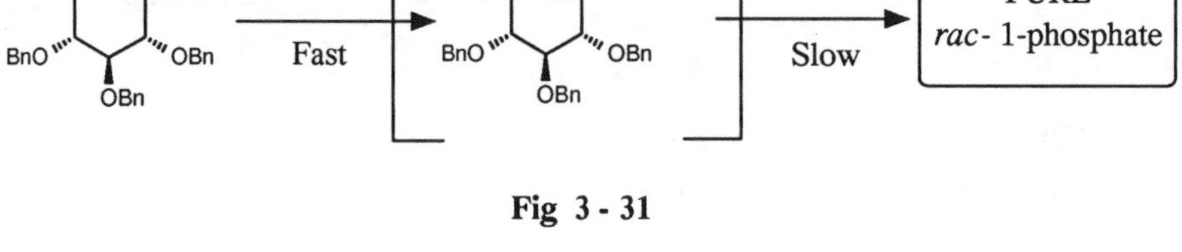

Fig 3 - 31

benzylphosphate esters are cleaved to the free phosphates much more rapidly than the benzyl ethers are cleaved to the free hydroxyl groups, thus avoiding the presence of free hydroxyl groups adjacent to phosphate esters during deprotection, *Fig 3-31*. This approach also has the advantage of allowing rigorous purification of the fully protected inositol phosphate by conventional chromatography on silica gel, followed by a single hydrogenolytic deprotection step, which normally leads directly to a pure product.

The problems of low reactivity of the inositol hydroxyls and formation of cyclic phosphates from *vic* diols have been solved by two general approaches. The first synthetic strategy involves the reaction of tetrabenzylpyrophosphate (TBPP) with alkoxide polyanions, *Fig 3-32*.[13,37,38] TBPP is a readily available, stable, crystalline reagent of relatively low toxicity, which presents no handling difficulties. This method gives very good results for polyhydroxy

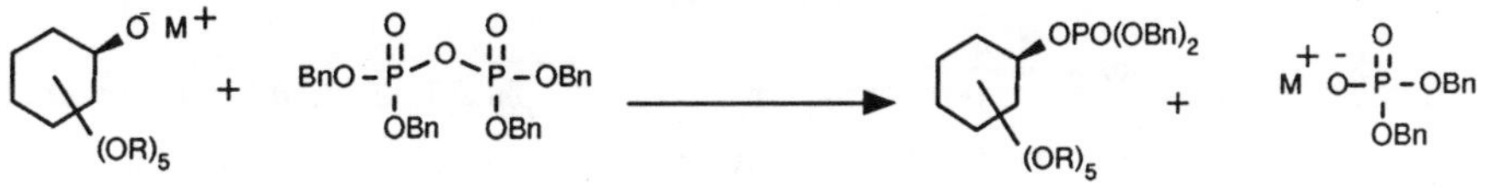

Fig 3 - 32

compounds, incorporating up to 4 protected phosphate groups in a single reaction. *vic* Diols are efficiently phosphorylated, without the formation of cyclic phosphate side products, and the desirable benzylphosphate protecting groups are introduced directly. An example of the use of this reagent is provided by the direct single-step phosphorylation of tetrol (20) giving fully protected 1,3,4,5-IP$_4$ (75) in 70% isolated yield, *Fig 3-33*, which on hydrogenolysis gives racemic 1,3,4,5-IP$_4$ (76) with no observed phosphate migration.[13,14] A number of variations on this strategy have been reported, involving the use of different bases and solvents for the alkoxide formation and subsequent *in-situ* phosphorylation reactions.

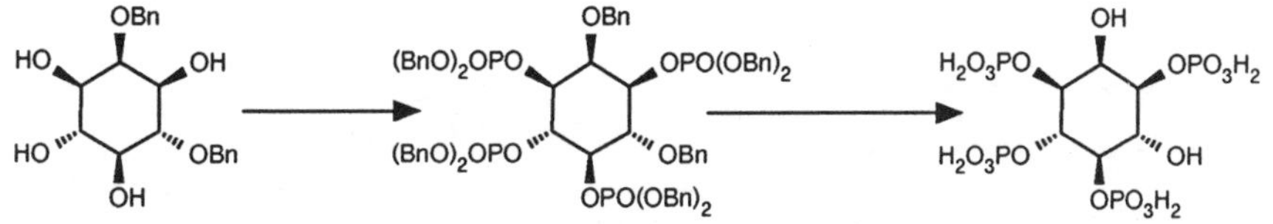

Fig 3 - 33

The second synthetic strategy involves the reaction of alcohols with P(III) reagents, well known to be much more reactive than their P(V) counterparts, followed by oxidation of the resulting phosphites to phosphates and subsequent deprotection, *Fig 3-34*.[39] This approach is based on the P(III) reagents originally developed for the synthesis of DNA.[40] Numerous variations on this theme have been used, involving the displacement of amine or halogen leaving groups from the P(III) reagent, variations of the oxidizing agent used to convert P(III)

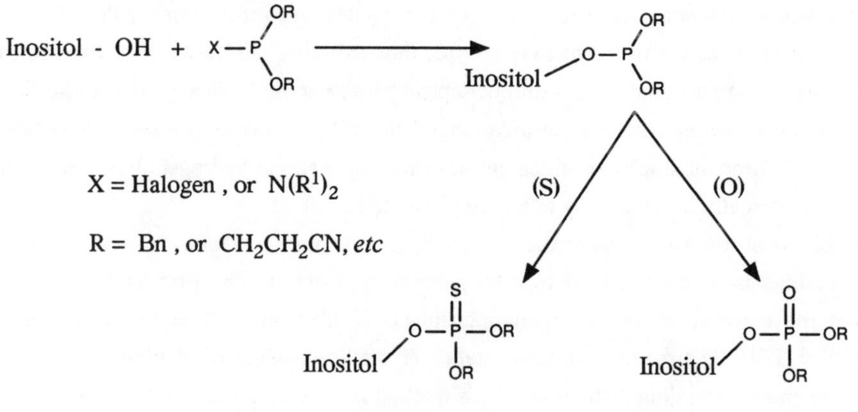

Fig 3 - 34

to P(V), and numerous different protecting groups for the phosphite / phosphate esters. This approach is also valuable as treatment of the intermediate phosphites with sulphur can give the phosphorothioate esters,[41] interesting analogues of the natural phosphates, but suffers from the relatively poor availability of many of the phosphite reagents. Probably the most useful variants of this approach are those which allow the introduction of benzyl or related

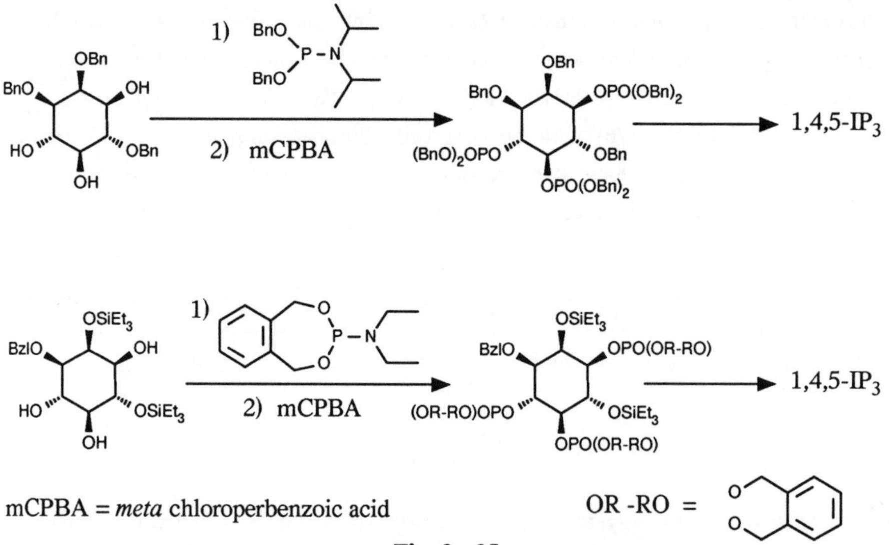

mCPBA = *meta* chloroperbenzoic acid

Fig 3 - 35

protecting groups directly at the phosphite stage, and two examples are given in *Fig 3-35*.[6,28,42] The efficiency of the phosphorylations now possible by this approach is illustrated by the greater than 90% isolated yields reported for both of the trisphosphorylations shown in *Fig 3-35*. The O-xylidine group shown in *Fig 3-35* is cleaved by hydrogenolysis under essentially the same conditions as a benzyl group.

References

1)   S.J.Angyal, M.E.Tate and S.D.Gero, *J.Chem.Soc.*, 1961, 4116.

2)   R.Gigg and C.D.Warren, *J.Chem.Soc.*(C), 1969, 2367.

3)   D.E.Kiely, G.J.Abruscato and V.Baburao, *Carbohydr.Res.*, 1974, **34**, 307.

4)   T.W.Greene and P.G.M.Wuts, "Protective Groups in Organic Synthesis", 2nd Edition, 1991, Wiley, New York.

5)   P.J.Garegg, T.Iversen, R.Johansson and B.Lindberg, *Carbohydr.Res.*, 1984, **130**, 322 and references therein.

6)   K.-L. Yu and B.Frazer-Reid, *Tetrahedron Lett.,* 1988, **29**, 979.

7)   N.Nagashima and N.Ohno, *Chem.Lett.*, 1987, 141.

8)   J.Gigg, R.Gigg, S.Payne and R.Conant, *Carbohydrate Res.,* 1985, **140**, c1-c3.

9)   Y.Watanabe, T.Ogasawara, H.Nakahira, T.Matsuki and S.Ozaki, *Tetrahedron Lett.,* 1988, **29**, 5259.

10)   J.Gigg, R.Gigg, S.Payne and R.Conant, *Carbohydrate Res.,* 1985, **142**, 132.

11)   C.B.Reese and J.G.Ward, *Tetrahedron Lett.,* 1987, **28**, 2309.

12)   H.W.Lee and Y.Kishi, *J.Org.Chem.*, 1985, **50**, 4402.

13)   D.C.Billington and R.Baker, *J.Chem.Soc.Chem.Commun.,* 1987, 1011.

14)   D.C.Billington, R.Baker, J.J.Kulagowski, I.M.Mawer, J.P.Vacca, S.J.de Solms and J.R.Huff, *J.Chem.Soc.Perkin Trans.1,* 1989, 1423.

15)   Y.Watanabe, M.Mitani, T.Morita and S.Ozaki, *J.Chem.Soc.Chem.Commun.*, 1989, 482.

16)   Y.Watanabe, T.Shinohara, T.Fujimoto and S.Ozaki, *Chem.Pharm.Bull.*, 1990, **38**, 562.

17)   A.Zapata, R.Fernandez de la Pradilla, M.Martin-Lomas and S.Penades, *J.Org.Chem.*, 1991, **56**, 444.

18)   I.H.Gilbert, A.B.Holmes and R.C.Young, *Tetrahedron Lett.,* 1990, **31**, 2633.

19)   S.V.Ley and F.Sternfeld, *Tetrahedron Lett.,* 1988, **29**, 5305.

20)   S.V.Ley, M.Parra, A.J.Redgrave and F.Sternfeld, *Tetrahedron,* 1990, **46**, 4995.

21)   H.A.J.Carless and K.Busia, *Tetrahedron Lett.*, 1990, **31**, 3449.

22)   J.G.Molotkovsky and L.D.Bergelson, *Tetrahedron Lett.*, 1971, 4791.

23)   J.G.Molotkovsky and L.D.Bergelson, *Chem.Phys.Lipids*, 1973, 135.

24)   Y-C.Liu and C-S.Chen, *Tetrahedron Lett.*, 1989, **30**, 1617.

25)   G.Baudin, B.I.Glanzer, K.S.Swaminathan and A.Vasella, *Helv.Chim.Acta.*, 1988, **71**, 1367.

26) D.C.Billington, *Chem.Soc.Rev.,* 1989, **18**, 83; A.E.Stepanov and V.I.Shvets, *Chem.Phys.Lipids,* 1979, **25**, 247.

27) S.Ozaki, Y.Watanabe, T.Ogasawara, Y.Kondo, N.Shiotani, H.Nishii and T.Matsuki, *Tetrahedron Lett.,* 1986, **27**, 3157.

28) Y.Watanabe, A.Oka, Y.Shimizu and S.Ozaki, *Tetrahedron Lett.,* 1990, **31**, 2613; Y.Watanabe, T.Fujimoto, T.Shinohara and S.Ozaki, *J.Chem.Soc.Chem.Commun.,* 1991, 428.

29) D.C.Billington, R.Baker, J.J.Kulagowski and I.M.Mawer, *J.Chem.Soc.Chem. Commun.,* 1987, 314.

30) R.Baker, J.J.Kulagowski, D.C.Billington, P.D.Leeson, I.C.Lennon and N.Liverton, *J.Chem. Soc.Chem.Commun.,* 1989, 1383; R.Baker, P.D.Leeson, N.Liverton and J.J.Kulagowski, *J.Chem.Soc.Chem. Commun.,* 1990, 462.

31) J.Gigg, R.Gigg, S.Payne and R.Conant, *J.Chem.Soc.Perkin Trans. 1,* 1987, 1757.

32) T.Desai, A.Fernandes-Mayoralas, J.Gigg, R.Gigg, C.Jaramilo, S.Payne, S.Penades and N.Schnetz, *ACS Symp.Ser.463*, "Inositol Phosphates and Derivatives", A.B.Reitz (Ed.), American Chemical Society, Washington, DC, 1991, 86.

33) C.E.Ballou and L.I.Pizer, *J.Am.Chem.Soc.,* 1960, **82**, 3333.

34) S.D.Gero, D.Mercier and J.E.G.Barnett, *Methods Carbohydrate Chem.,* 1972, **6**, 403; D.Mercier, J.E.G.Barnett and S.D.Gero, *Tetrahedron,* 1969, **25**, 5681; S.D.Gero, *Tetrahedron Lett.,* 1966, 591; D.Mercier and S.D.Gero, *Tetrahedron Lett.,* 1968, 3459.

35) W.Tegge and C.E.Ballou, *Proc.Natl.Acad.Sci.USA,* 1989, **86**, 94.

36) J.R.Falck and P.Yadagiri, *J.Org.Chem.,* 1989, **54**, 5851.

37) Y.Watanabe, H.Nakahira, M.Bunya and S.Ozaki, *Tetrahedron Lett.,* 1987, **28**, 4179.

38) J.P.Vacca, S.J.de Solms and J.R.Huff, *J.Am.Chem.Soc.,* 1987, **109**, 3478.

39) A.M.Cooke, R.Gigg and B.V.L.Potter, *Biochem.Soc.Trans,* 1987, **15**, 904; M.R.Hamblin, B.V.L.Potter and R.Gigg, *J.Chem.Soc.Chem.Commun.,* 1987, 626.

40) N.D.Sinha, J.Biernat, J.McManus and H.Koster, *Nucleic Acids Res.,* 1984, **12**, 4539.

41) A.M.Cooke, R.Gigg and B.V.L.Potter, *J.Chem.Soc.Chem.Commun.,* 1987, 1525.

42) Y.Watanabe, Y.Komoda, K.Ebisuya and S.Ozaki, *Tetrahedron Lett.,* 1990, **31**, 255.

Chapter 4

Synthesis of Inositol Monophosphates

4.1 Introduction

Four isomeric *myo*-inositol monophosphates exist, and all have been synthesised. Due to the previously discussed symmetry of the *myo*-inositol molecule, inositol 1-phosphate and inositol 4-phosphate exist as pairs of enantiomers, whilst inositol 2-phosphate and inositol 5-phosphate are *meso* compounds (plane of symmetry through C-2 and C-5). In addition to the synthesis of racemic (*rac*) mixtures, both of the optically active monophosphates have been synthesised as individual enantiomers.

4.2 Inositol 1-Phosphate

4.2.1 Syntheses of Racemic Material

Hydrolysis of natural phospholipids with alkali leads to a dextrorotatory inositol monophosphate. The structure of this monophosphate was determined to be D-(+)-inositol

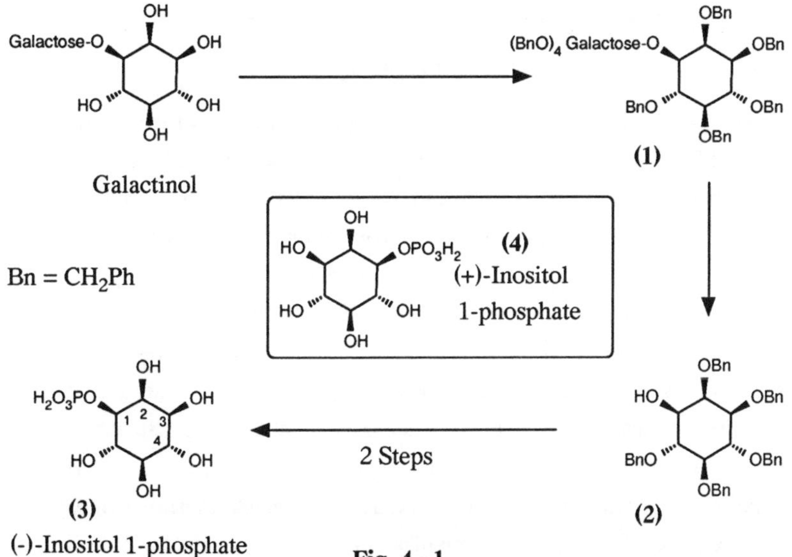

Fig 4 - 1

1-phosphate by Pizer and Ballou in 1959.[1] The absolute configuration of this material was established [2] by synthesis of its enantiomer, L-(-)-inositol 1-phosphate, from galactinol of known absolute configuration, *Fig 4-1*. Perbenzylation of natural galactinol gave the nonabenzyl derivative (1), which gave the pentabenzyl inositol (2) on hydrolysis of the galactose moiety. Phosphorylation of the free hydroxyl group, followed by cleavage of the benzyl protecting groups gave a laevorotatory phosphate L-(-)-inositol 1-phosphate (3). The naturally occurring isomer of inositol 1-phosphate isolated from phospholipids is therefore the enantiomer of this synthetic compound *ie* (4). The isolation of the naturally occurring enantiomer (4) from phospholipids on a preparative scale is possible,[3] if rather tedious. It is not possible to obtain the enantiomeric material (3) from natural sources.

Early syntheses of inositol 1-phosphate were unselective, involving the separation of

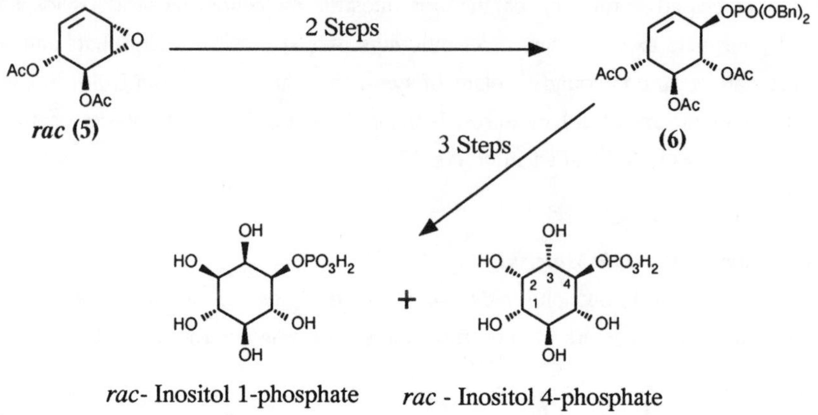

rac - Inositol 1-phosphate   *rac* - Inositol 4-phosphate

Fig 4 - 2

isomeric phosphates as a final step. The first reported synthesis [4] of *rac*-inositol 1-phosphate used diacetyl-1,2-anhydroconduritol (5) as starting material, *Fig 4-2*. Epoxide opening using dibenzylphosphate as nucleophile, followed by acetylation of the resulting free hydroxyl group gave the protected phosphate (6). Hydroxylation of the double bond in (6) gave a mixture of the two possible *cis* diols [attack from above or below the plane of the ring as drawn in (6)]. Deprotection (2 steps) of this mixture of diols gave a mixture of *rac*-inositol 1-phosphate and *rac*-inositol 4-phosphate. These isomeric phosphates were separated by crystallisation to give pure *rac*-inositol 1-phosphate [*ie* a mixture of (3) and (4)] in very low recovery. A similar separation of isomers was required [5] after the phosphorylation of the diol (8) derived in two steps from the previously described monoacetal (7), *Fig 4-3*. Direct phosphorylation of (8) with $POCl_3$ gave a mixture of protected inositol 1-phosphate (10) and inositol 2-phosphate (9). Careful crystallisation led to the isolation of *rac* (10), and deprotection gave *rac*-inositol 1-phosphate.

A more selective approach was used by Gigg and Warren [6] who exploited the higher reactivity of the equatorial hydroxyl group in the diol (8), *Fig 4-4*. Reaction of (8) with allyl bromide under carefully controlled conditions gave mainly the equatorial O-allyl compound (11). Benzylation of the free hydroxyl group, followed by removal of the allyl group, led to the desired selectively protected inositol (12). Phosphorylation with the highly reactive

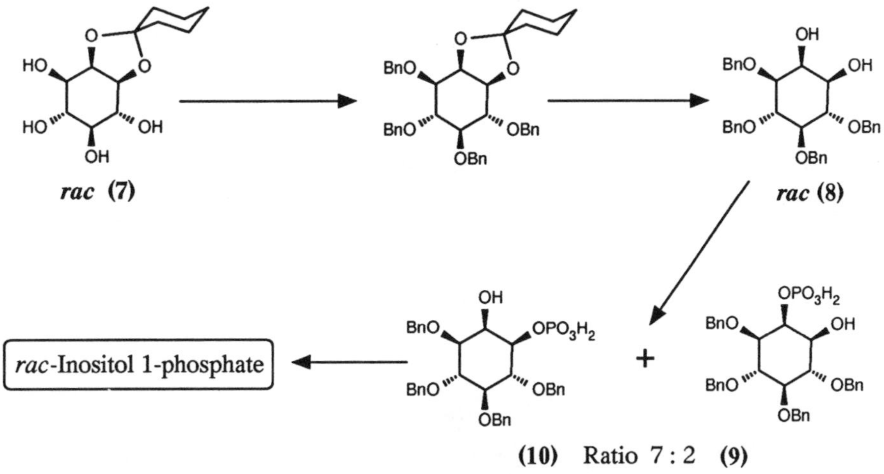

Fig 4 - 3

diphenylchlorophosphate then gave the pentabenzyl diphenylphosphate (13). Deprotection by hydrogenolysis, first with a palladium catalyst to remove the benzyl ethers, followed by a platinum catalyst to remove the phenyl phosphate esters produced a 98:2 mixture of *rac*-inositol 1-phosphate and inositol 2-phosphate (14). This mixture arises *via* formation and

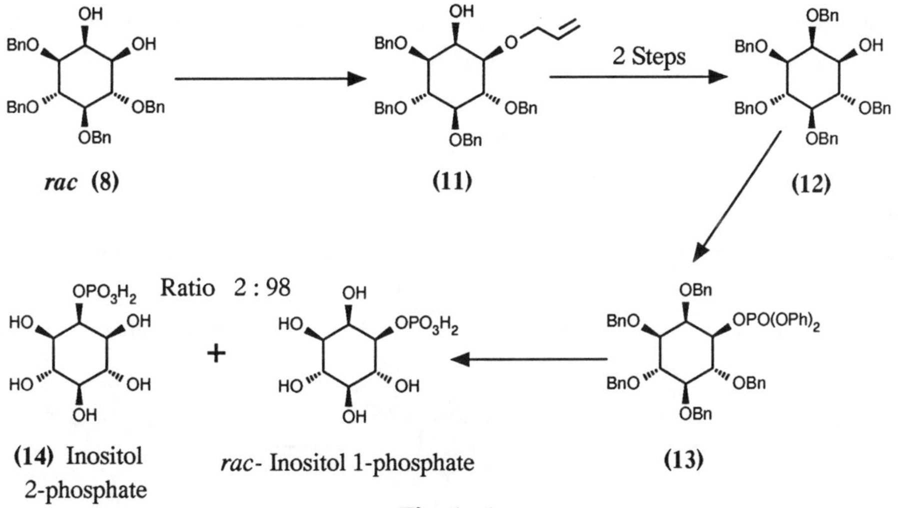

Fig 4 - 4

subsequent unselective opening of a cyclic phenyl phosphate ester (15) formed from the intermediate diphenyl phosphate (16) produced in the two-step deprotection procedure,

Fig 4-5. A general solution to this problem is to use benzyl esters as phosphate protecting groups, in place of phenyl esters.[7,8] Transesterification of the fully protected compound (13)

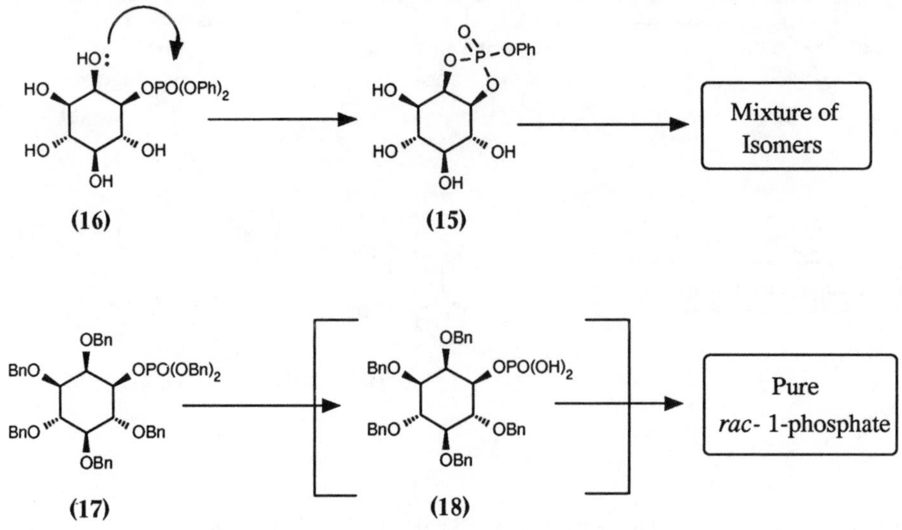

(16)   (15)   Mixture of Isomers

(17)   (18)   Pure rac- 1-phosphate

Fig 4 - 5

using the anion of benzyl alcohol [7] gives the dibenzyl phosphate (17). This ester may then be deprotected by a single hydrogenolysis over a palladium catalyst,[7] as the benzyl phosphate esters are cleaved much faster than the benzyl ethers, giving a free phosphate (18) which is not prone to migration during the slower hydrogenolysis of the benzyl ethers, *Fig 4-5*. This use of benzyl ethers in conjunction with benzyl phosphate esters has been subsequently applied in the synthesis of many inositol phosphates and analogues, without migration of phosphate groups occurring.

In a more classical approach, the pentaacetate (19)[9,10,11] was treated directly with $POCl_3$ to give the crude phosphorylated material (20),[12] *Fig 4-6*. Preparative HPLC allowed the isolation of the pure phosphate (20). Crude (20) could be deprotected to give *rac*-inositol

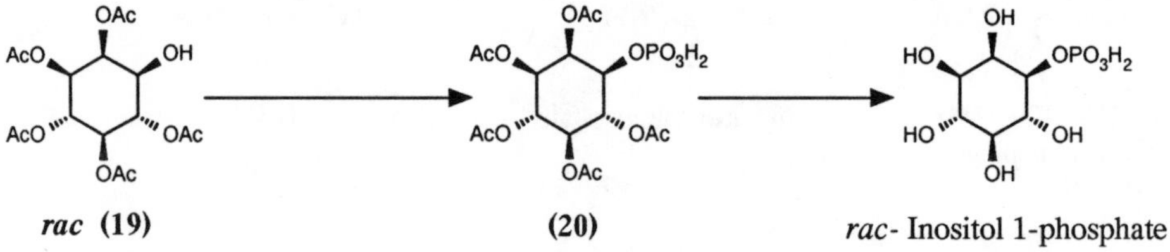

rac (19)   (20)   *rac*- Inositol 1-phosphate

Fig 4 - 6

1-phosphate, which was isolated in pure form by crystallisation.

The fundamental approach of Ley *et al* to the synthesis of inositol phosphates has been outlined in Chapter 3, and is exemplified, *Fig 4-7*, by their synthesis of *rac*-inositol 1-phosphate.[13] The diol (21) was converted in 7 steps to the epoxide (22) as described previously (Chapter 3). MM2 calculations suggested that the preferred conformation of (22) was a well-defined boat, and the NMR coupling constants of the ring protons supported this suggestion. This led to the conclusion that regioselective epoxide opening should occur *via* preferential attack at C-6, and treatment of (22) with the anion derived from alcohol (23) led to the protected *myo*-inositol derivative (24). Phosphorylation of (24) with tetrabenzylpyrophosphate followed by hydrogenolysis of the benzyl esters and ethers gave the diol phosphate (26). Final deprotection of (26) with trifluoroacetic acid led to *rac*-inositol 1-phosphate in low yield.

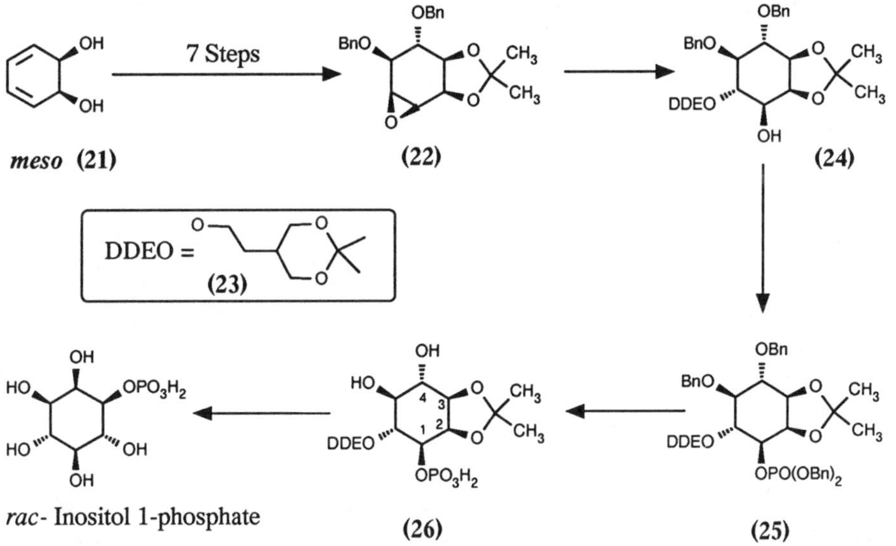

Fig 4-7

4.2.2 (-) and (+)-Inositol 1-Phosphate

The synthesis of (-)-inositol 1-phosphate (3) from galactinol has been described in *Fig 4-1*. The optically active natural product quebrachitol (27) has been used as a starting point for the synthesis of both (-) and (+)-inositol 1-phosphate (3) and (4).[14-18] Treatment of the natural product (27) with cyclohexanone gives the bisacetal [15] (28), a common intermediate in the synthesis of both (3) and (4).

Towards a synthesis of (-)-inositol 1-phosphate (3), [14-17] tosylation of (28) gave the tosylate (29), and treatment with boron trichloride cleaved both the acetal and methyl protecting groups to give pentol (30), *Fig 4-8*. Benzoylation of this pentol gave the fully protected

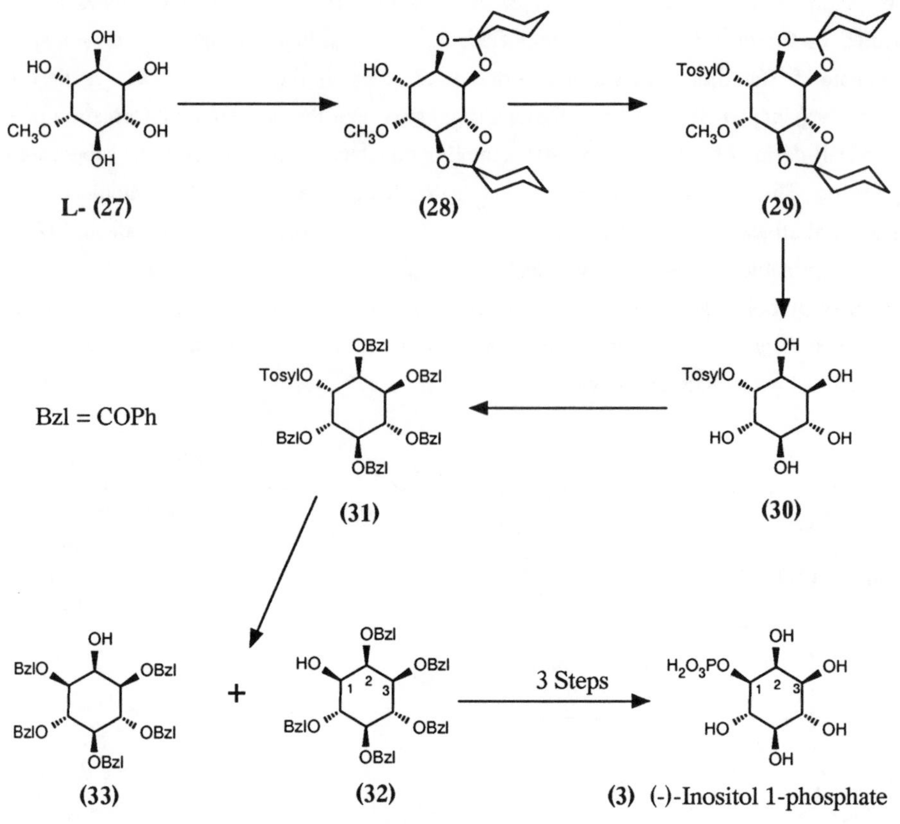

Fig 4 - 8

inositol (31), which on treatment with NaF in N,N-dimethylformamide underwent intramolecular benzoyloxy displacement of the tosyl ester, giving a mixture of the desired optically active 2,3,4,5,6-pentabenzoyl inositol (32) and the isomeric *meso* material (33).[14-17] Recrystallisation afforded pure (32), which could be phosphorylated using diphenyl-chlorophosphate. Deprotection by hydrogenolysis of the phenyl phosphate esters, followed by basic hydrolysis of the benzoyl esters gave pure (-)-inositol 1-phosphate (3).

The synthesis of (+)-inositol 1-phosphate involves oxidation of alcohol (28) to the ketone, followed by stereoselective reduction with $LiBH_4$ to give the epimeric alcohol (34), *Fig 4-9* , which was protected as the benzoate (35).[18] In this case attempted demethylation *via* BCl_3 and HI both caused problems, and only $NaI/AlCl_3$ was effective for clean selective demethylation, cleaving both the methyl and the *trans* acetal protecting groups to give (36). Perbenzoylation of (36), followed by acid hydrolysis of the *cis* acetal gave the diol (37).

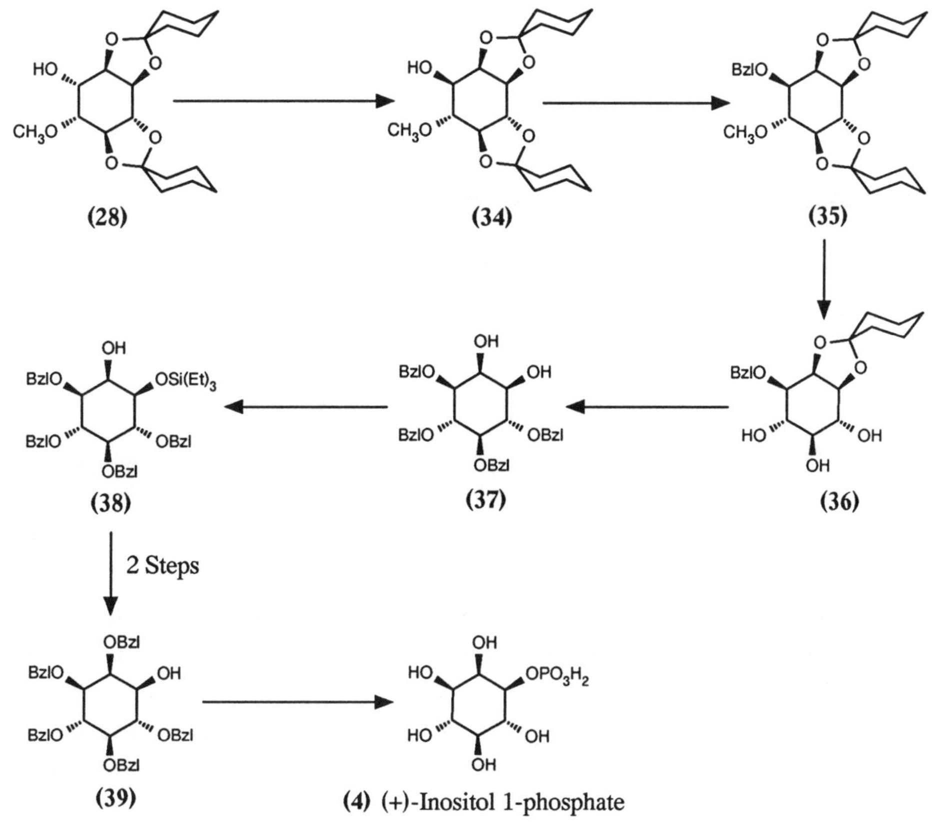

Fig 4 - 9

Highly selective silylation of (37) with triethylsilyl chloride gave exclusively the 1-silylated alcohol (38). Benzoylation and desilylation then gave the pentabenzoyl alcohol (39). Phosphorylation using the new reagent combination [19] 3-diethylamino-1,5-dihydro-2,4,3-benzodioxaphosphepine / mCPBA followed by hydrogenolysis and base-mediated debenzoylation gave (+)-inositol 1-phosphate (4).

A number of methods have been applied to the resolution of intermediates suitable for the preparation of the enantiomers of inositol 1-phosphate.

Conversion of the racemic pentaacetate (19) into its acid oxalate gives (40), *Fig 4-10*, which may be resolved in the classical manner *via* salt formation with chiral bases.[20,21] Unfortunately, to obtain good yields of resolved material it is necessary to use quinidine to obtain one isomer and (-)-α-phenylethylamine to obtain the other. In addition multiple crystallisations are required to obtain the diastereomeric salts in pure form. The enantiomeric pentaacetates obtained (41) and (42) were transformed into the enantiomers of inositol

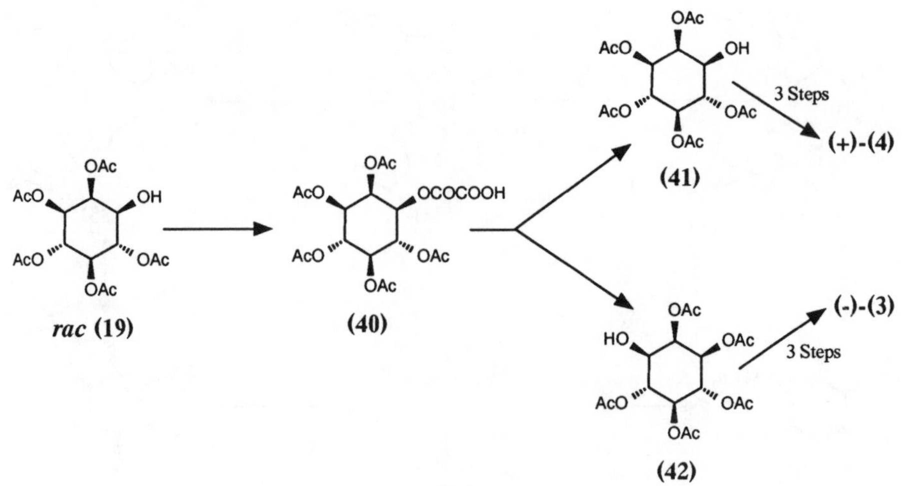

Fig 4 - 10

1-phosphate (3) and (4) by standard methods.

Conversion of racemic inositol derivatives into diastereomeric mixed orthoesters has been used to resolve a number of useful intermediates.[22,23] Treatment of the racemic pentabenzyl inositol (12) with the orthoester of D-mannose gave the diastereomeric mixed orthoesters (43) and (44) , *Fig 4-11*, which were separated by a combination of chromatography on alumina and crystallisation.[23] Mild acid hydrolysis then regenerated the pure enantiomers (45) and (46).

 In a more recent approach,[7] the alcohol (12) has also been resolved by conversion into its diastereomeric camphanate esters, (47) and (48) *via* treatment with commercially available (-)-camphanic acid chloride. The camphanates were easily separated by chromatography on silica, and the diastereomeric purity of the esters determined by HPLC. Single crystal X-ray structure analysis of one of the camphanate esters allowed the absolute configuration of the esters, and thus of the inositol phosphates derived from them, to be determined. As the absolute configuration of the camphanic acid used was known, this analysis involved no stereochemical assumptions, and confirmed the original assignment of structure (4) to the dextrorotatory isomer of inositol 1-phosphate. Phosphorylation of (45) and (46) with diphenylchlorophosphate, followed by transesterification with the anion of benzyl alcohol (see Section 4.2.1 and *Fig 4.5*) gave the dibenzylphosphates, which were deprotected by a single hydrogenolysis to give the pure enantiomers of inositol 1-phosphate (3) and (4).

This use of camphanate esters has been discussed in detail in Chapter 3, and the approach has now been successfully used for the resolution of a wide range of protected inositol derivatives. The major advantages of conventional chromatographic separation of diastereomers, and HPLC determination of diastereomeric purity (and thus enantiomeric purity), coupled with the commercial availability of (-)-camphanic acid chloride make this the first choice method for the resolution of new inositol derivatives.

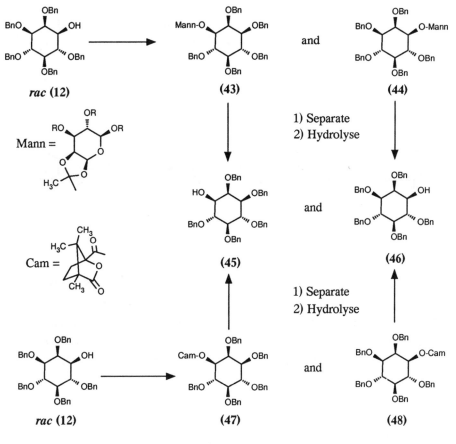

Fig 4 - 11

4.3 Inositol 2-Phosphate

An optically inactive inositol monophosphate was isolated from the acidic hydrolysis of wheat bran in 1912, and subsequently shown to be inositol 2-phosphate (14).[24,25] It has since become clear that the penultimate product of the acidic, alkaline, or enzymic hydrolysis of inositol hexakisphosphate (phytic acid) is inositol 2-phosphate.

The first reported synthesis, *Fig 4-12*, of the 2-phosphate took advantage of the highly selective oxidation of *myo*-inositol at the 2-position by *Acetobacter suboxydans* giving *scyllo*-inose (49).[24] Acetylation of (49) followed by reduction of the ketone by hydrogenolysis gave the protected *myo*-inositol (50).[25] Phosphorylation with diphenyl-chlorophosphate and subsequent deprotection then gave inositol 2-phosphate (14).

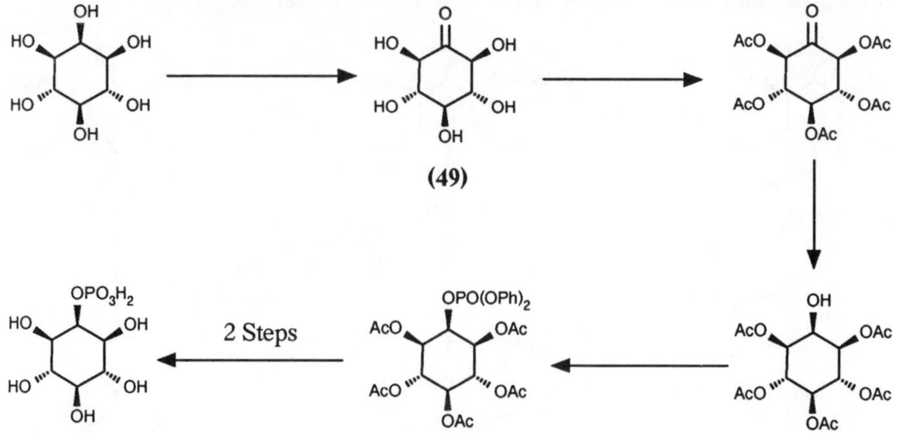

Fig 4 - 12

A more conventional approach used the previously described selective alkylation of the tetrabenzyl diol (8).[26] Benzylation of (8) gives predominantly the pentabenzyl alcohol (52), *Fig 4-13*. Phosphorylation with diphenylchlorophosphate under forcing conditions, followed by transesterification leads to the fully protected compound (53). A single hydrogenolysis then removes all of the protecting groups, giving pure inositol 2-phosphate (14).

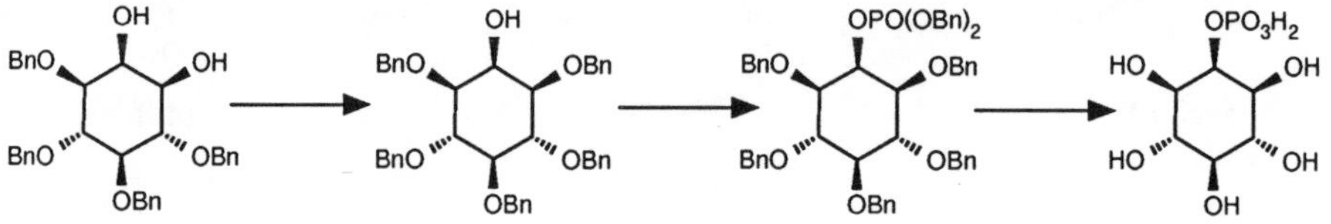

Fig 4 - 13

The previously described orthoformate (54) (see Chapter 3) may be selectively dibenzylated at the 4- and 6- positions by taking advantage of the chelation-controlled alkylations of mono anions of (54).[27] Thus, *Fig 4-14*, sequential treatment of (54) with 1 equiv. sodium hydride and benzyl bromide, followed , after alkylation is complete, by a second equivalent of base and alkylating agent, leads to the dibenzyl intermediate (55) in low yield. Phosphorylation of (55) with sodium hydride / tetrabenzylpyrophosphate gave the fully protected compound

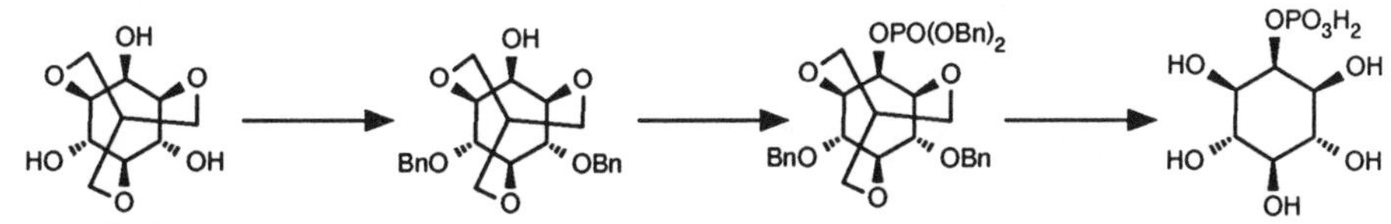

(54) (55) (56) (14)

Fig 4 - 14

(56), and deprotection by hydrogenolysis followed by acidic hydrolysis of the orthoformate gave pure inositol 2-phosphate (14).

4.4 Inositol 4-Phosphate

In addition to the dextrorotatory 1-phosphate, alkaline hydrolysis of brain-derived phospholipids gives minor amounts of a second optically active inositol monophosphate, and this is therefore one enantiomer of inositol 4-phosphate.[28]

4.4.1 Syntheses of Racemic Material

The majority of reported racemic syntheses use the previously described (Chapter 3) biscyclohexylidine acetal (57) as a key intermediate. Selective reactions at the more reactive 3-hydroxyl group of (57) are possible, and the monobenzoate (58a),[29] monobenzyl ether (58b)[23] and monomannose orthoester (58c)[7] have all been prepared and used successfully in

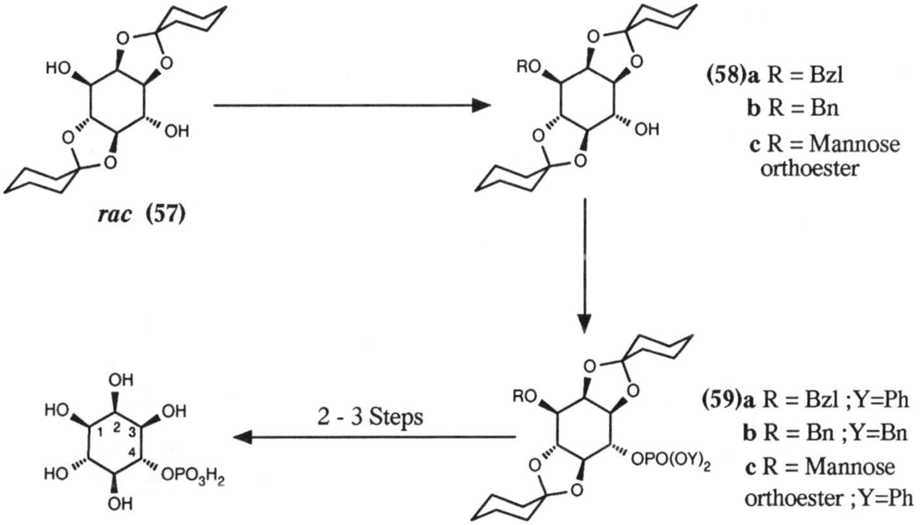

rac (57)

(58)a R = Bzl
b R = Bn
c R = Mannose orthoester

(59)a R = Bzl ;Y=Ph
b R = Bn ;Y=Bn
c R = Mannose orthoester ;Y=Ph

rac-Inositol 4-phosphate

Fig 4 - 15

synthesis, *Fig 4-15*. Phosphorylation of the remaining free hydroxyl group in (58a-c) gives the corresponding fully protected compounds (59a-c).

As C-4 and C-6 of *myo*-inositol are enantiotopic (see Chapter 1, Section 1.2.2) deprotection of (59a-c) by suitable methodology led to *rac*-inositol 4-phosphate in each case, thus exploiting the inherent symmetry of the parent system.

A more recent, and much shorter, route[8,27] uses the highly selective chelation-controlled phosphorylation of the mono anion of (54) to produce the protected phosphate (60) in a single step (*ie* in only two steps from inositol itself). Deprotection by hydrogenolysis of the benzyl phosphate esters, followed by an acidic work-up to cleave the orthoformate, gave *rac*-inositol 4-phosphate in excellent yield.

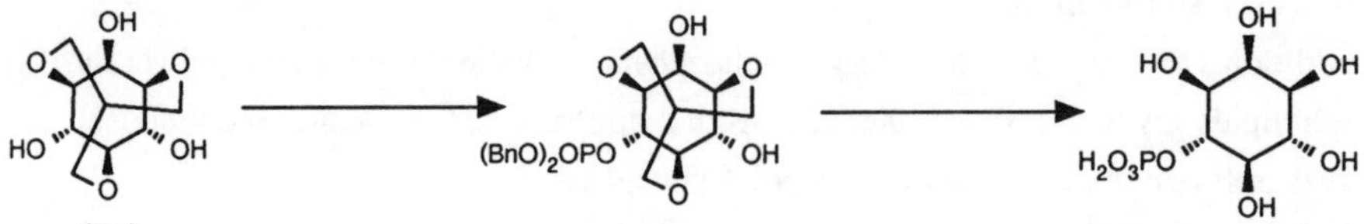

(54)   (60)   *rac*-Inositol 4-phosphate

Fig 4 - 16

Reaction of the acetal (7) with 1,3-dichloro-1,1,3,3- tetraisopropyldisiloxane (TIPS) results in selective formation of the diol (61), which may be selectively benzoylated to give the alcohol (62), *Fig 4-17*.[30] Treatment of (62) with tetrabenzylpyrophosphate and base resulted in migration of the benzoyl group and phosphorylation giving the fully protected phosphate (63). Sequential deprotection gave *rac*-inositol 4-phosphate in good yield.

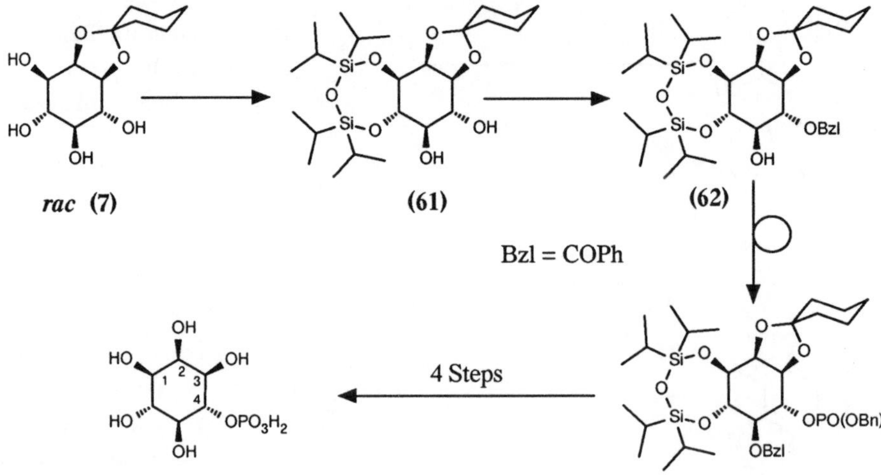

rac (7)   (61)   (62)

Bzl = COPh

4 Steps

rac-Inositol 4-Phosphate   (63)

Fig 4 - 17

4.4.2 (+) and (-)-Inositol 4-Phosphate

Attempts to separate the diastereomeric mannose orthoesters of (57), *ie* the diastereomers of (58c), have not succeeded to date.[23] In contrast,[7] treatment of (58b) with (-)-camphanic acid chloride, *Fig 4-18*, gives a mixture of readily separable camphanate esters (64) and (65). Conventional chromatography on silica and crystallisation yielded the pure esters, whose diastereomeric purity could be determined by HPLC. Hydrolysis of the camphanate esters gave the free enantiomers of (58b), which were converted into the enantiomers of inositol 4-phosphate, (66) and (67), by the same method used in the racemic series.[7]

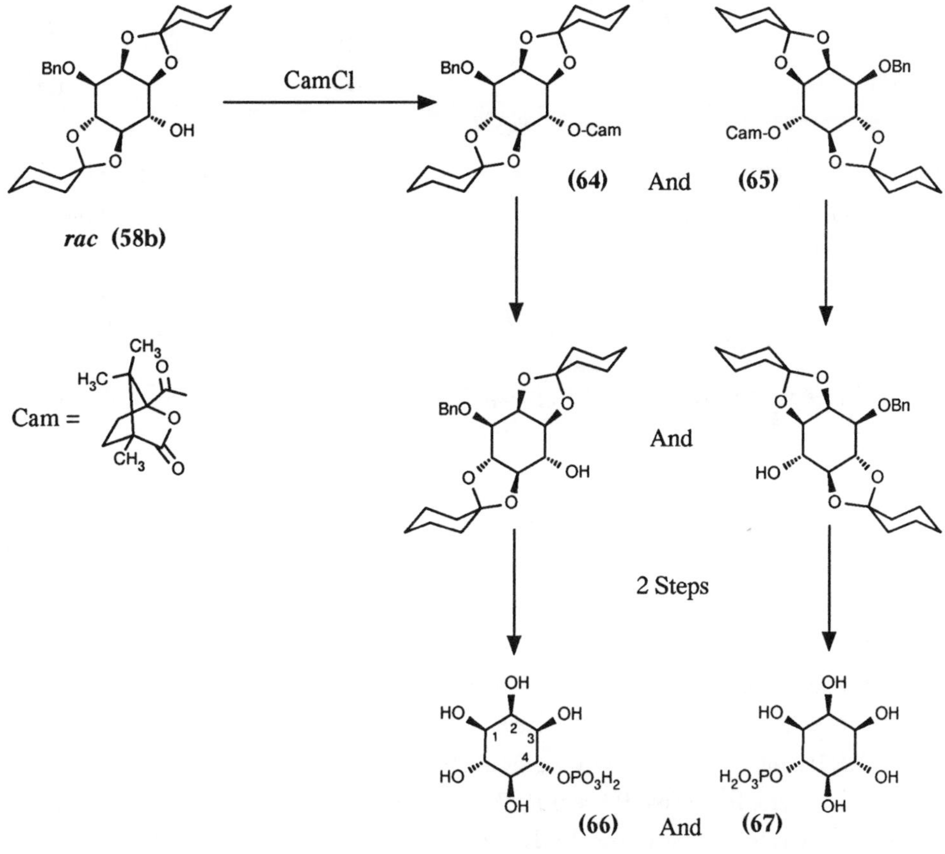

(+) and (-) Inositol 4-phosphate

Fig 4 - 18

The absolute configuration of the inositol 4-phosphates has not been confirmed by physical means to date.

4.5 Inositol 5-Phosphate

The first reported synthesis of inositol 5-phosphate,[29] started from the rather inaccessible material 2-amino-2-deoxy-*neo*-inositol (68), *Fig 4-19*. In this study, the starting material (68) was obtained either by hydrolysis of the antibiotic hygromycin A,[31] or by synthesis.[32] Protection of (68) as its pentaacetate (69) allowed nitrous acid oxidation of the amino function, giving the protected *myo*-inositol (70). Phosphorylation with diphenylchloro-phosphate, followed by deprotection gave inositol 5-phosphate [29] (71).

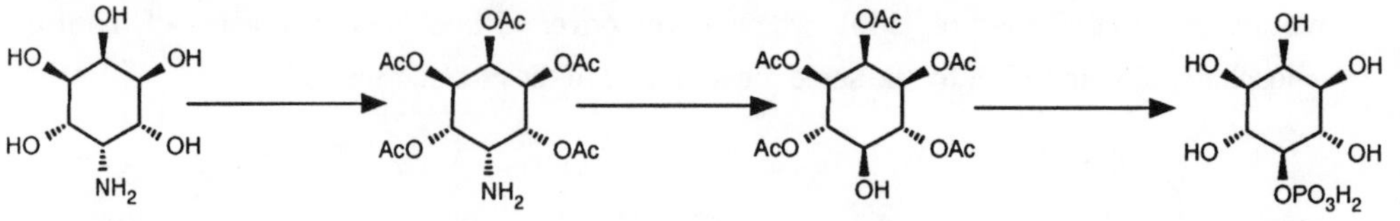

Fig 4 - 19

The protected TIPS derivative (62) (see *Fig 4-17*) may also be converted into inositol 5-phosphate, *Fig 4-20*, by phosphitylation using PCl_3 and benzyl alcohol, followed by oxidation to the protected phosphate and subsequent deprotection.[30]

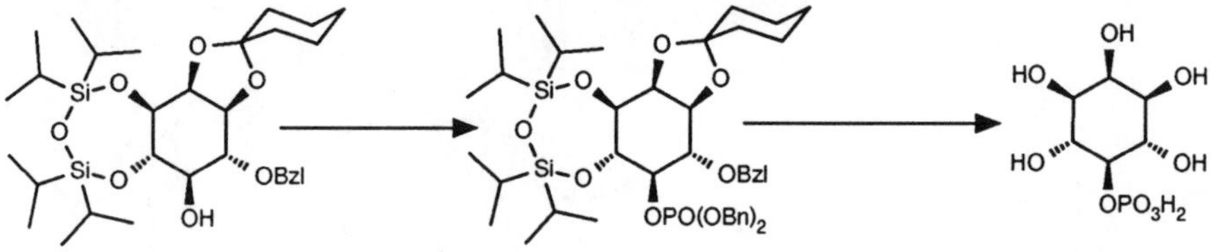

Fig 4 - 20

References

1)   L.I.Pizer and C.E.Ballou, *J.Am.Chem.Soc.*, 1959, **81**, 915.

2)   C.E.Ballou and L.I.Pizer, *J.Am.Chem.Soc.*, 1960, **82**, 3333.

3)   C.E.Ballou, *Biochemical Preparations*, 1962, **9**, 99.

4)   N.Kurihara, H.Shibata, H.Saeki and M.Nakajima, *Liebigs Ann.Chem.*, 1967, **701**, 225.

5)   D.E.Kiely, G.J.Abruscato and V.Baburao, *Carbohydrate Res.*, 1974, **34**, 307.

6)   R.Gigg and C.D.Warren, *J.Chem.Soc. (C)*, 1969, 2367.

7)   D.C.Billington, R.Baker, J.J.Kulagowski and I.M.Mawer, *J.Chem.Soc.Chem. Commun.*, 1987, 314.

8)   D.C.Billington and R.Baker, *J.Chem.Soc.Chem.Commun.*, 1987 1011.

9)   S.J.Angyal, M.E.Tate and S.D.Gero, *J.Chem.Soc.*, 1961, 4116.

10) S.J.Angyal and M.E.Tate, *J.Chem.Soc.*, 1965, 6949.

11) S.J.Angyal, M.H.Randall and M.E.Tate, *J.Chem.Soc.(C)*, 1967, 919.

12) T.Metschies, C.Schultz and B.Jastorff, *Tetrahedron Lett.*, 1988, **29**, 3921.

13) S.V.Ley, M.Parra, A.J.Redgrave and F.Sternfeld, *Tetrahedron*, 1990, **46**, 4995.

14) S.D.Gero, D.Mercier and J.E.G.Barnett, *Methods Carbohydrate Chem.*, 1972, **6**, 403.

15) D.Mercier, J.E.G.Barnett and S.D.Gero, *Tetrahedron*, 1969, **25**, 5681.

16) S.D.Gero, *Tetrahedron Lett.*, 1966, 591.

17) D.Mercier and S.D.Gero, *Tetrahedron Lett.*, 1968, 3459.

18) T.Akiyama, N.Takechi and S.Ozaki, *Tetrahedron Lett.*, 1990, **31**, 1433.

19) Y.Watanabe, Y.Komoda, K.Ebisuya and S.Ozaki, *Tetrahedron Lett.*, 1990, **31**, 255.

20) J.G.Molotkovsky and L.D.Bergelson, *Tetrahedron Lett.*, 1971, 4791.

21) J.G.Molotkovsky and L.D.Bergelson, *Chem.Phys.Lipids*, 1973, 135.

22) V.I.Shvets, B.A.Klyaschitskii, A.E.Stepanov and R.P.Evstigneeva, *Tetrahedron*, 1973, **29**, 331.

23) A.E.Stepanov, O.O.Tutorskaya, B.A.Klyaschitskii, V.I.Shvets and R.P.Evstigneeva, *Zh.Obs.Chim.*, 1972, **42**, 709; S.P.Kozlova, I.S.Pekarskaya, B.A.Klyaschitskii, V.I.Shvets and R.P.Evstigneeva, *Zh.Obs.Chim.*, 1972, **42**, 702; B.A.Klyaschitskii, V.V.Pimenova, A.I.Bashkatova, E.G.Zhelvakova, S.D.Sokolov, V.I.Shvets and R.P.Evstigneeva, *Zh.Obs.Chim.*, 1970, **40**, 2482; Reviewed in : A.E.Stepanov and V.I.Shvets, *Chem.Phys.Lipids*, 1979, **25**, 247.

24) T.Posternak, *Helv.Chim.Acta*, 1941, **24**, 1045.

25) B.M.Iselin, *J.Am.Chem.Soc.*, 1949, **71**, 3822.

26) D.C.Billington and J.J.Kulagowski, *unpublished observations*.

27) D.C.Billington, R.Baker, J.J.Kulagowski, I.M.Mawer, J.P.Vacca, S.J.de Solms and J.R.Huff, *J.Chem.Soc.Perkin Trans.1*, 1989, 1423.

28) C.Grado and C.E.Ballou, *J.Biol.Chem.*, 1961, **236**, 54.

29) S.J.Angyal and M.E.Tate, *J.Chem.Soc.*, 1961, 4112.

30) Y.Watanabe, M.Mitani, T.Morita and S.Ozaki, *J.Chem.Soc.Chem.Commun.*, 1989, 482.

31) J.B.Patrick, R.P.Williams, C.W.Waller and B.L.Hutchins, *J.Am.Chem.Soc.*, 1956, **78**, 2652; R.L.Mann and D.O.Woolf, *J.Am.Chem.Soc.*, 1957, **79**, 120.

32) G.R.Allen, *J.Am.Chem.Soc.*, 1956, **78**, 5691.

Chapter 5

Synthesis of Inositol Bisphosphates

5.1 Introduction

Base-catalysed hydrolysis of brain-derived phosphoinositides gives a mixture of inositol bisphosphates, from which inositol 1,4- and 4,5-bisphosphates were isolated in the early 1960s.[1,2] Inositol 1,2-bisphosphate has been isolated from the products obtained by treatment of phytic acid (inositol hexakisphosphate) with phytase enzymes derived from both microbial and plant sources.[3,4,5] Inositol 1,3-, 1,4- and 3,4-bisphosphates have all been confirmed as intermediates in the recycling of inositol phosphates *via* the PI cycle,[6] see Chapter 2. Syntheses of all of these molecules, except the 1,2- isomer, have been reported.

5.2 *meso*-Bisphosphates ; Inositol 1,3-Bisphosphate

The only naturally occurring *meso*-bisphosphate to be identified to date is inositol 1,3-bisphosphate, an intermediate in the PI cycle (see Chapter 2).

The synthesis of this material has been reported, by a route that takes advantage of the symmetry of the inositol orthoformate derivative (1) (see Chapter 3).[7] Exhaustive benzylation

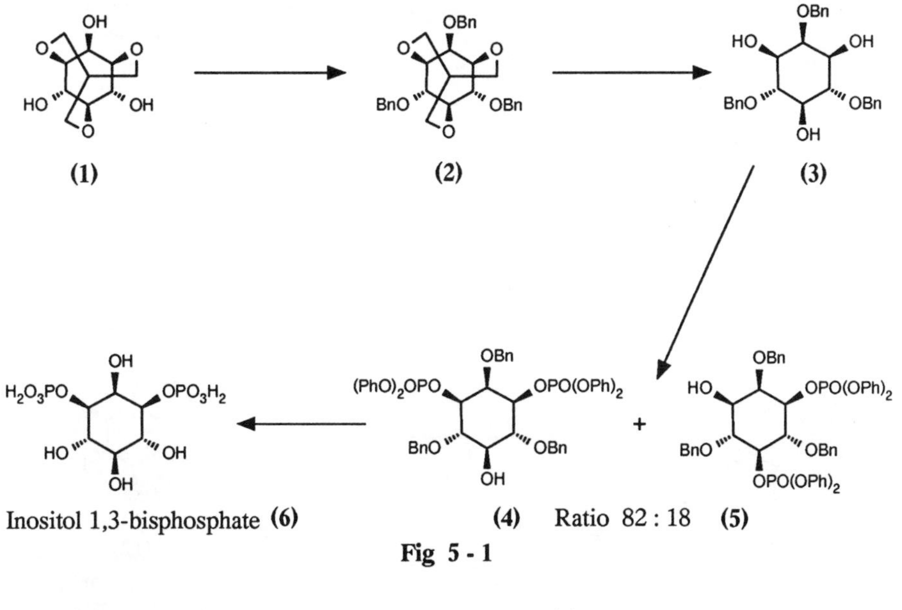

Inositol 1,3-bisphosphate **(6)**   **(4)**   Ratio 82 : 18   **(5)**

Fig 5 - 1

of the orthoformate (1) gives the fully protected inositol (2), which on acidic hydrolysis gives the tribenzyl inositol (3),[7,8] *Fig 5-1*. The 1- and 3-hydroxyl groups of triol (3) each have one axial and one equatorial neighbour, and are thus less hindered than the 5-hydroxyl group, which has two equatorial neighbours. Phosphorylation of (3) with diphenylchlorophosphate proceeds with reasonable selectivity to give an 82:18 mixture of 1,3- and 1,5-bisphosphorylated products, (4) and (5). Benzylation of (3) gave predominantly the expected 1,2,4,6-tetra-O-benzyl product, under a variety of conditions, for similar reasons. Fortunately the desired 1,3-bisphosphorylated material (4) can be isolated in a pure state by crystallisation.

Attempts to transesterify the phenyl phosphates to benzyl phosphates failed in this case. Deprotection of (4) was finally achieved directly by careful treatment with lithium in liquid ammonia / THF at -78°C,[9] to give the desired inositol 1,3-bisphosphate (6).

5.3 Optically Active Bisphosphates
5.3.1 *Naturally Occurring Isomers ; Racemic Syntheses*

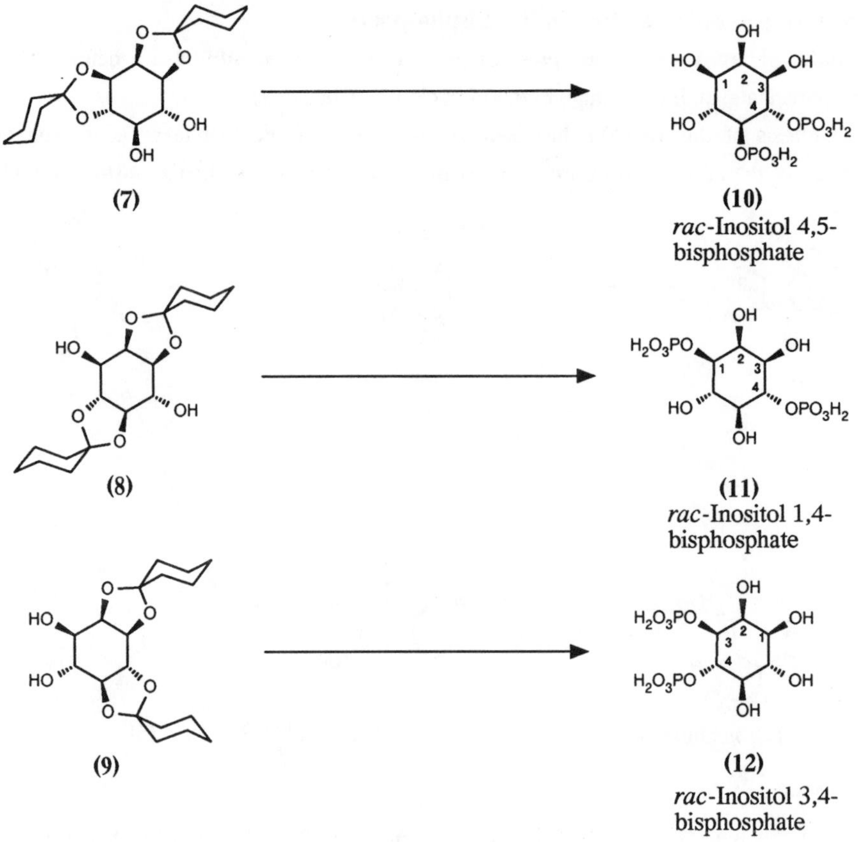

(7)

(10)
rac-Inositol 4,5-
bisphosphate

(8)

(11)
rac-Inositol 1,4-
bisphosphate

(9)

(12)
rac-Inositol 3,4-
bisphosphate

Fig 5 - 2

The most direct route to the 1,4- , 3,4- and 4,5-bisphosphates is *via* direct phosphorylation of the three isomeric bisacetals, (7), (8) and (9), whose chemistry has been discussed in Chapter 3. Syntheses using this approach were reported in 1961 by Angyal and Tate,[10] *Fig 5-2*. The original approach used diphenylchlorophosphate as phosphorylating agent, but a more recent report using the phosphite / oxidation technology has appeared, and this is probably more efficient overall.[11] This phosphite approach was also used for the phosphorylation of tetrabenzyl inositols *eg* (13) to give access to the same bisphosphates, *eg Fig 5-3*.[12,13]

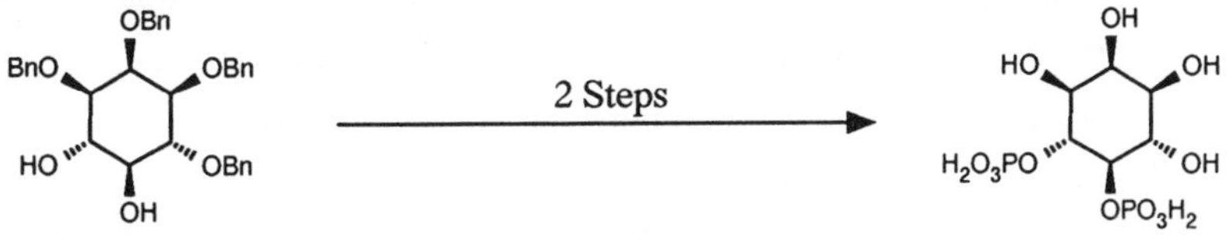

Fig 5 - 3

A much less direct approach[14] illustrates the use of conduritol B derivatives such as (18) for the synthesis of *myo*-inositol phosphates, *Fig 5-4*. This strategy is discussed in general in Chapter 3, and although the use of this approach for bisphosphates seems rather contrived, it has been used to provide viable routes to inositol tris- and tetrakisphosphates (see Chapters 6 and 7).

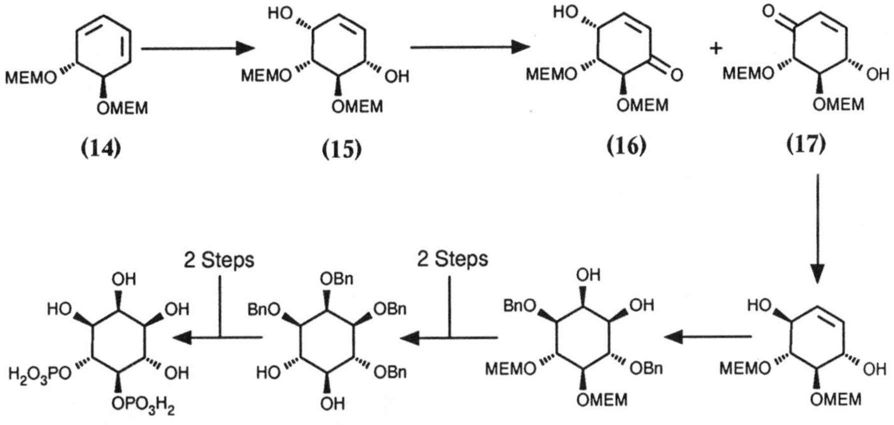

Fig 5 - 4

[4 + 2] Addition of singlet oxygen to the protected *trans* diol (14) produces the protected tetrol (15). Oxidation of (15) to a separable mixture of (16) and (17), followed by reduction of the desired isomer (17) with $NaBH_4/CeCl_3$ then gave the diol (18). Benzylation of (18) followed by *cis* hydroxylation using the catalytic OsO_4 system then produced the protected *myo*-inositol derivative (19). Benzylation of (19), followed by removal of the MEM protecting groups finally gave diol (20), which on phosphorylation with tetrabenzylpyrophosphate and deprotection by hydrogenolysis led to *rac*-inositol 4,5-bisphosphate (10).

5.3.2 Naturally Occurring Isomers ; Syntheses of Individual Enantiomers

The individual enantiomers of these three bisphosphates have all been synthesised from the corresponding enantiomers of the bisacetals,[15] obtained by resolutions using the orthoesters of D-mannose.[16,17] The use of these orthoesters of D-mannose has been previously described for the resolution of intermediates in the synthesis of inositol 1-phosphate (Chapter 4) and a review of this resolution method is available.[18] Phosphorylation of the resolved bisacetals was performed using diphenylchlorophosphate in the classical manner, and hydrogenolysis gave the enantiomeric bisphosphates.

A more recent approach[19] uses the conversion of the bisacetal (8) into its diastereomeric biscamphanate esters (21) and (22) *via* treatment with 2 equiv. of (-)-camphanic acid chloride,[20] *Fig 5-5*. These diesters were easily separated by chromatography, and hydrolysis gave the enantiomeric bisacetals (23) and (24). Phosphorylation with diphenylchlorophosphate and deprotection by hydrogenolysis then gave the individual enantiomers of inositol 1,4-bisphosphate.

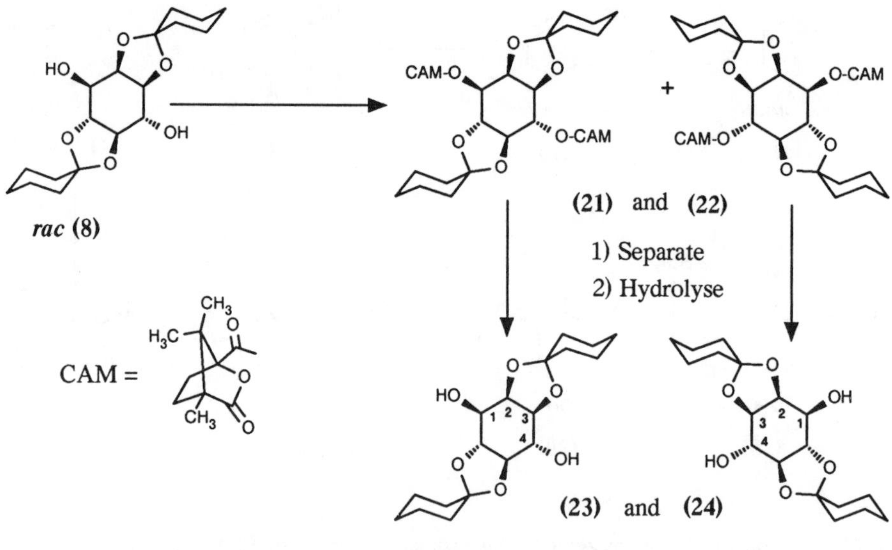

Fig 5 - 5

This approach is interesting as it illustrates the concept that a protected inositol derivative having *N* free hydroxyl groups can in theory, steric hindrance *etc* notwithstanding, be resolved by conversion into a pair of fully protected diastereomeric camphanate esters having *N* camphanate groups by treatment with > *N* equiv. of (-)-camphanic acid chloride.

5.3.3 Non-Naturally Occurring Isomers

In studies aimed at elucidating the structure-activity relationships involved in calcium mobilisation by D-inositol 1,4,5-trisphosphate, both enantiomers of inositol 1,5-bisphosphate have been synthesised.[21] Perbenzylation of the orthoformate (1) followed by cleavage of the orthoformate gave the triol (3), as before, *Fig 5-1*. Benzylation of (3) under phase transfer conditions led to predominantly the expected 1,2,4,6-tetra-O-benzyl inositol (25), *via* reaction at the less hindered hydroxyl group, *Fig 5-6* (*cf* the phosphorylation reaction in *Fig 5-1*). Treatment of (25) with 1.1 equiv. of (-)-camphanic acid chloride led to the diastereomeric camphanates (26) and (27) by selective ester formation at the less hindered hydroxyl group.

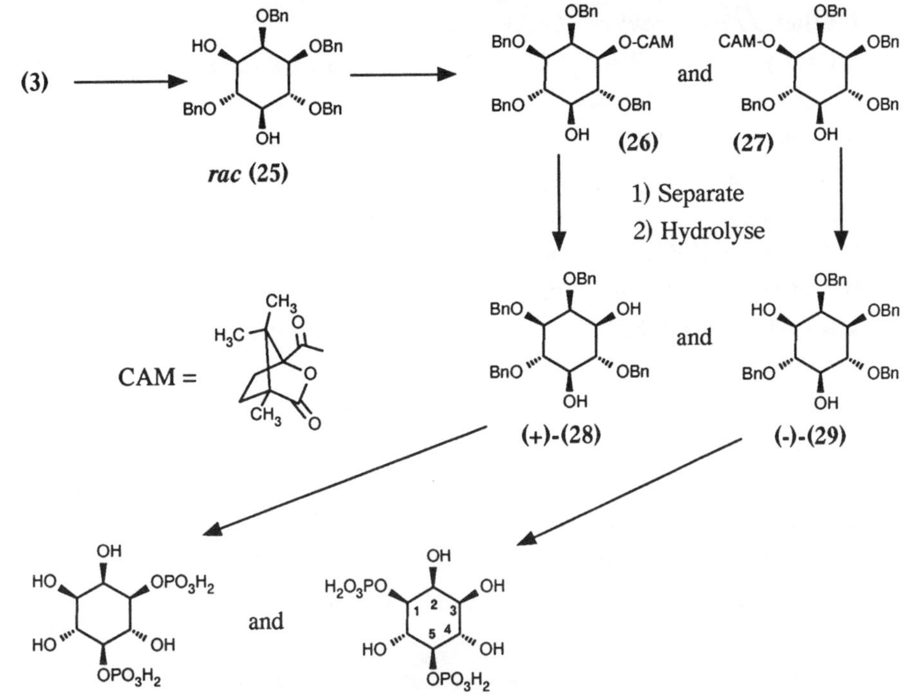

Fig 5 - 6

The diastereomeric esters were separated by conventional chromatography, and hydrolysis gave the enantiomerically pure diols (+)-(28) and (-)-(29). Phosphorylation *via* a phosphoramidite reagent,[22] followed by oxidation to the phosphate level and deprotection gave the individual enantiomers of inositol 1,5-bisphosphate (+)-(30) and (-)-(31).

References

1)   C.Grado and C.E.Ballou, *J.Biol.Chem.*, 1961, **236**, 54.

2)   R.V.Tomlinson and C.E.Ballou, *J.Biol.Chem.*, 1961, **236**, 1902.

3)   L.F.Johnson and M.E.Tate, *Ann.N.Y.Acad.Sci.*, 1969, **165**, 526.

4)   R.V.Tomlinson and C.E.Ballou, *Biochemistry*, 1962, **1**, 166.

5)   P.E.Lim and M.E.Tate, *Biochim.Biophys.Acta*, 1973, **302**, 316.

6)   B.V.L.Potter, Chapter 11.4 in "Comprehensive Medicinal Chemistry", 1990, Pergamon Press, Oxford.

7)   D.C.Billington and R.Baker, *J.Chem.Soc.Chem.Commun.*, 1987, 1011.

8)   D.C.Billington, R.Baker, J.J.Kulagowski, I.M.Mawer, J.P.Vacca, S. J.deSolms and J.R.Huff, *J.Chem.Soc.Perkin Trans.1*, 1989, 1423.

9)   B.E.Maryanoff, A.B.Reitz, G.F.Tutwiler, S.J.Benkovic, P.A.Benkovic and S.J.Pilkis, *J.Am.Chem.Soc.*, 1984, **106**, 7851.

10)   S.Angyal and M.E.Tate, *J.Chem.Soc.*, 1961, 4122.

11)   M.R.Hamblin, J.S.Flora and B.V.L.Potter, *Biochem.J.*, 1987, **246**, 771.

12)   M.R.Hamblin, B.V.L.Potter and R.Gigg, *J.Chem.Soc.Chem.Commun.*, 1987, 626.

13)   M.R.Hamblin, B.V.L.Potter and R.Gigg, *Biochem.Soc.Trans.*, 1987, **15**, 415.

14)   H.A.J.Carless and J.Busia, *Tetrahedron Lett.*, 1990, **31**, 3449.

15)   V.N.Krylova, N.I.Kobel'kova, G.F.Oleinik and V.I.Shvets, *Zh.Org.Khim.*, 1980, **16**, 62.

16)   V.I.Shvets, B.A.Klyaschitskii, A.E.Stepanov and R.P.Evstigneeva, *Tetrahedron*, 1973, **29**, 331.

17)   A.E.Stepanov, O.O.Tutorskaya, B.A.Klyaschitskii, V.I.Shvets and R.P.Evstigneeva, *Zh.Obs.Chim.*, 1972, **42**, 709; S.P.Kozlova, I.S.Pekarskaya, B.A.Klyaschitskii, V.I.Shvets and R.P.Evstigneeva, *Zh.Obs.Chim.*, 1972, **42**, 702; B.A.Klyaschitskii, V.V.Pimenova, A.I.Bashkatova, E.G.Zhelvakova, S.D.Sokolov, V.I.Shvets, R.P.Evstigneeva and N.A.Preobrazhenskii, *Zh.Obs.Chim.*, 1970, **40**, 2482.

18)   A.E.Stepanov and V.I.Shvets, *Chem.Phys.Lipids*, 1979, **25**, 247.

19)   J.P.Vacca, S. J.deSolms, J.R.Huff, D.C.Billington, R.Baker, J.J.Kulagowski and I.M.Mawer, *Tetrahedron*, 1989, **45**, 5769.

20)   D.C.Billington, R.Baker, J.J.Kulagowski and I.M.Mawer, *J.Chem.Soc.Chem. Commun.*, 1987, 314.

21)   P.Westerduin, H.A.M.Willems and C.A.A.van Boeckel, *Tetrahedron Lett.*, 1990, **31**, 6915.

22)   C.B.Reese and J.G.Ward, *Tetrahedron Lett.*, 1987, **28**, 2309.

Chapter 6

Synthesis of Inositol Trisphosphates

6.1 Introduction

Due to the second messenger role of inositol 1,4,5-trisphosphate (1,4,5 -IP$_3$) outlined in Chapter 2, and the relative complexity of the molecule, the synthesis of this inositol phosphate and its isomers/analogues has been a topic of major interest in recent years. Following the identification of 1,4,5-IP$_3$ as the intracellular calcium mobilising signal in 1984, the first successful chemical synthesis of 1,4,5-IP$_3$ was reported in late 1986. Subsequent developments and diversification of synthetic approaches have produced many syntheses of this important natural molecule. The search for analogues of 1,4,5-IP$_3$ with altered pharmacological properties (resistance to metabolic degradation, antagonism of the 1,4,5-IP$_3$ response, *etc*) has led to the synthesis of a number of isomeric trisphosphates.

As the first reported selective synthesis of an inositol polyphosphate was the 1986 synthesis of D-(-)-1,4,5-IP$_3$, and as this has been followed by both racemic and chiral syntheses, a separation of the approaches reported under the headings racemic and chiral would lead to an incorrect impression of the chronology and the inter-relationships of the various syntheses. For this reason the syntheses of the inositol tris- and tetrakisphosphates will be presented in chronological order, arranged by substitution pattern.

6.2 Inositol 1,4,5-Trisphosphate

The first reported synthesis of 1,4,5-IP$_3$ by Ozaki *et al* exploited the higher reactivity of the equatorial *cf* axial OH group in protected inositol 1,2-diols on two occasions.[1] This selectivity is based on the work of Gigg and Warren reported almost twenty years earlier (see below).[2]

Exhaustive benzylation of the 1,2;4,5-bisacetal described previously (see Chapter3) gives the fully protected inositol (1), *Fig 6-1*. Selective hydrolysis of the less stable *trans* acetal, followed by allylation of the resulting diol gave the monoacetal (2), which on acidic hydrolysis gave the diol (3). Resolution of this racemic diol was accomplished by conversion to the diastereomeric monomenthoxyacetyl monoester derivatives (4) and (5) (see Chapter 3), ester formation occurring only at the more reactive equatorial position. The diastereomeric esters were separated by chromatography and crystallisation to give the desired optically pure intermediate ester (5). Hydrolysis of the ester (5) then produced (-)-(6), *ie* the desired enantiomer of (3). Selective allylation of the 1-OH group gave the alcohol (7), and this was followed by benzylation of the free 2-OH and cleavage of the three allyl groups to give the key optically active triol (+)-(8).

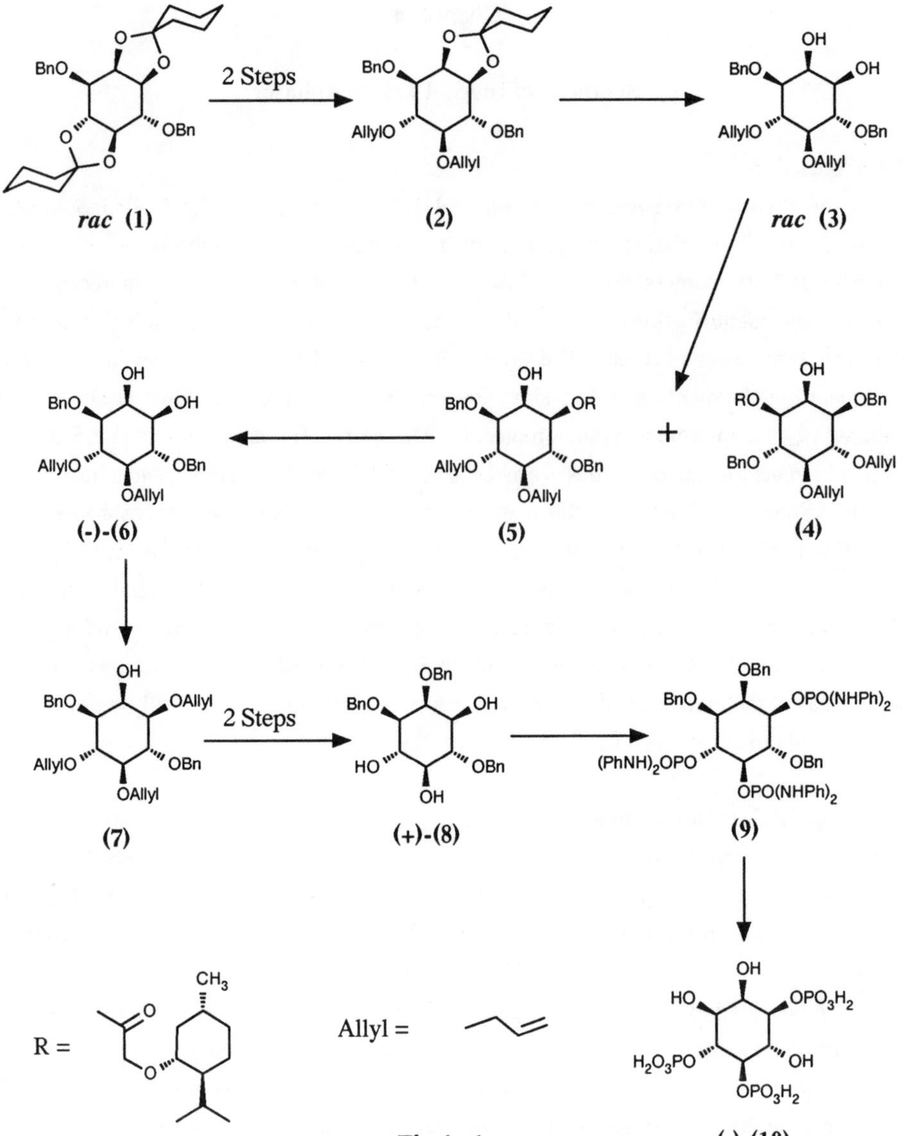

Fig 6 - 1

It should be emphasised at this point, that at the time that this synthesis was performed, the lack of suitable phosphorylation methods for polyhydroxy alcohols was a major stumbling block to chemists in the area. The triol (+)-(8) was phosphorylated with dianilido-chlorophosphate, to give the fully protected inositol trisphosphate (9). Sequential deprotection of (9) then gave the desired natural isomer D-(-)-inositol 1,4,5-trisphosphate (10), but due to the inefficiency of the phosphorylation/deprotection strategy only low yields of (-)-(10) could be obtained. A more efficient synthesis of (-)-(10) would result from the use of one of the more recently developed phosphorylation/deprotection strategies, *eg* phosphorylation of the alkoxide with tetrabenzylpyrophosphate,[3,4] phosphitylation followed by oxidation,[5] *etc*. These generally applicable strategies have been described in Chapter 3, and are used to advantage in the syntheses described below.

The severe problems associated with efficient phosphorylation of polyols at this time are underlined by the fact that Gigg's group reported [6] a synthesis of racemic triol (8) in 1985, but were unable to report the conversion of this material into racemic 1,4,5-IP$_3$ until 1987.[7]

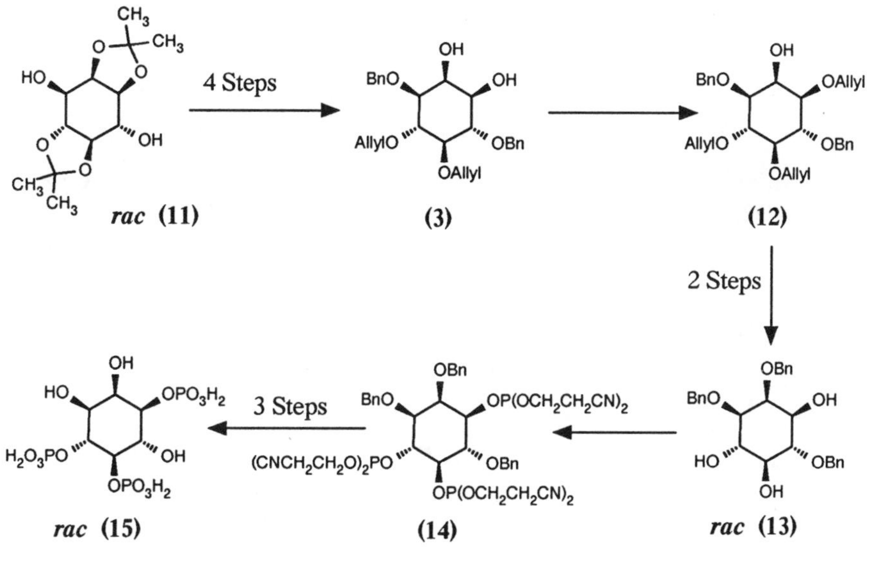

Fig 6 - 2

Starting from the bisisopropylidine acetal (11),[8] analogous to the cyclohexylidine acetal used by Ozaki's group, benzylation of the free OH groups, followed by acidic hydrolysis of the less stable *trans* acetal, allylation of the exposed OH groups, and hydrolysis of the *cis* acetal, gave the diol (3), *Fig 6-2*. Treatment of the diol (3) with tributyltin oxide,[9] and alkylation of the cyclic stannylene formed with allyl bromide gave the racemic tri-O-allyl derivative (12), *ie* the racemic form of (7), in high yield. Benzylation of the free OH followed by cleavage of the three allyl groups then gave the racemic triol (13). Reaction of (13) with ClP(OCH$_2$CH$_2$CN)N(CHMe$_2$)$_2$, followed by displacement of the diisopropylamine group with cyanoethanol gave the phosphite (14). Oxidation of the phosphite to the phosphate, followed by deprotection in a stepwise fashion gave *rac*-1,4,5-IP$_3$, (15).[7] Treatment of (14) with S$_8$ in pyridine followed by deprotection was used to obtain the trisphosphorothioate analogue of 1,4,5-IP$_3$.

In a related approach [10,11] selective protection of the more reactive 4-OH group of the

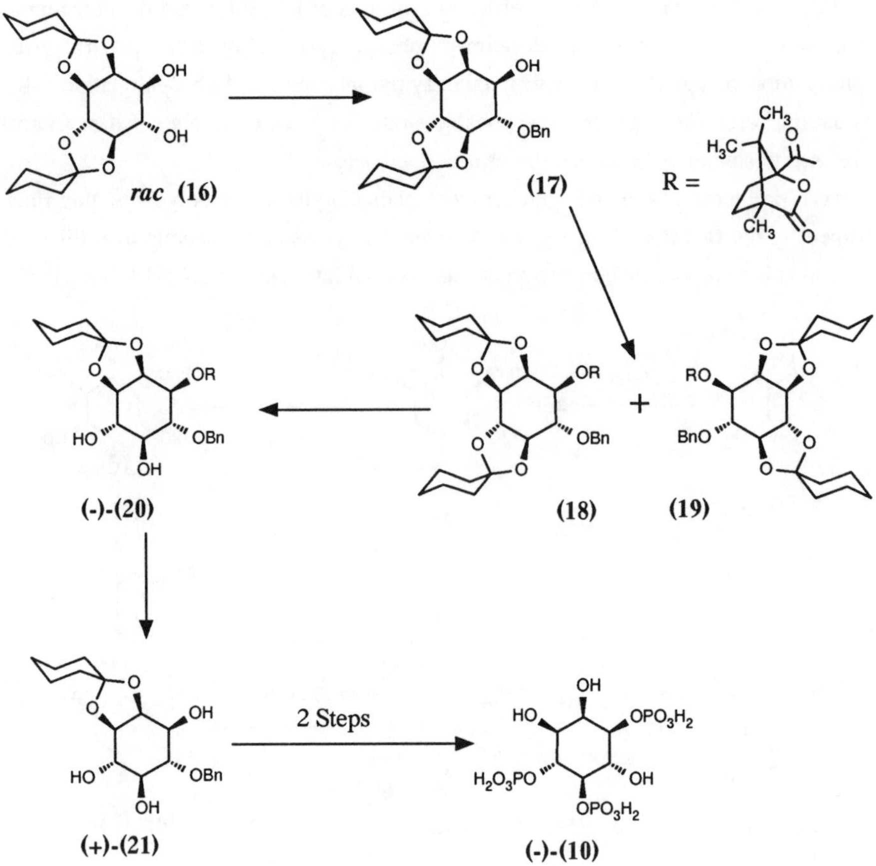

Fig 6 - 3

1,2;5,6-biscyclohexylidine acetal (16) as the benzyl ether gave the alcohol (17) in 40% yield, *Fig 6-3*. Conversion of this alcohol into its diastereomeric camphanate esters with (-)-camphanic acid chloride, followed by separation of the diastereomers (18) and (19), and hydrolysis of the less stable *trans* acetal of the desired isomer, gave the optically active diol (-)-(20). Basic hydrolysis of the camphanate ester then provided the triol (+)-(21). This triol was efficiently phosphorylated using KH/ THF/ tetrabenzylpyrophosphate to give the fully benzylated trisphosphate. Deprotection by hydrogenolysis, and acidic removal of the final acetal group then gave (-)-1,4,5-IP$_3$, (-)-(10), in excellent yield. Repetition of the above sequence using the other camphanate ester (19) provided the enantiomeric material, (+)-1,4,5-IP$_3$. The value of tetrabenzylpyrophosphate as a phosphorylating agent is underlined by the *ca* 60% overall yield of (-)-(10) from the triol (+)-(21).

A less conventional approach [12] uses the cyclopentylidine acetal (23) derived from the

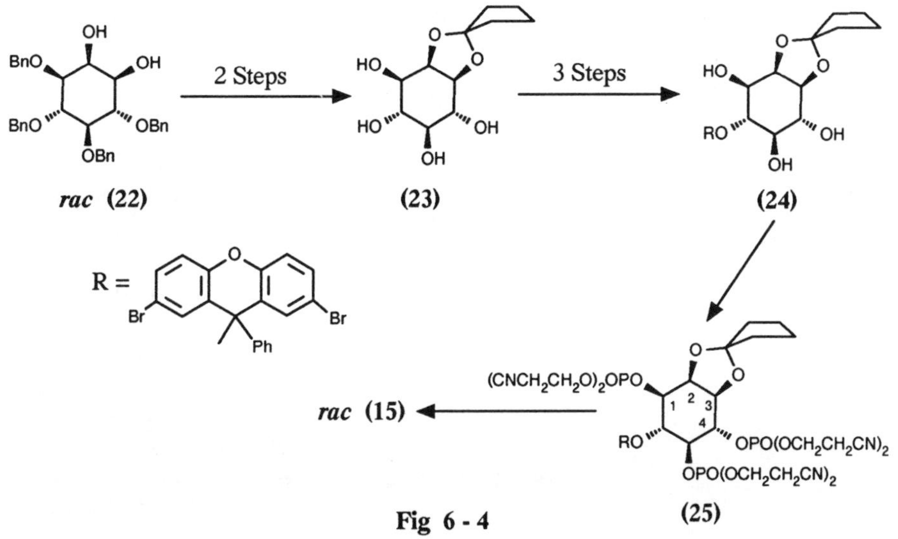

Fig 6 - 4

racemic tetrabenzyl diol (22), whose synthesis has been described previously (see Chapter 4), by acetal formation and removal of the benzyl ethers using Na/NH$_3$, *Fig 6-4*. In a three-step sequence the triol (24) is obtained in 30-40% overall yield by silylation of (23) with *t*-butyldimethylsilyl chloride, reaction of the crude product with 9-chloro-2,7-dibromo-9-phenylxanthene, and subsequent desilylation. Phosphitylation of this triol, and oxidation then gives the fully protected racemic phosphate (25). Deprotection in a stepwise manner gave *rac*-1,4,5-IP$_3$, (15). Repetition of the sequence using the enantiomerically pure diol derived from (22) by the resolution method of Stepanov [13] allowed synthesis of (-)-1,4,5-IP$_3$, (-)-(10).

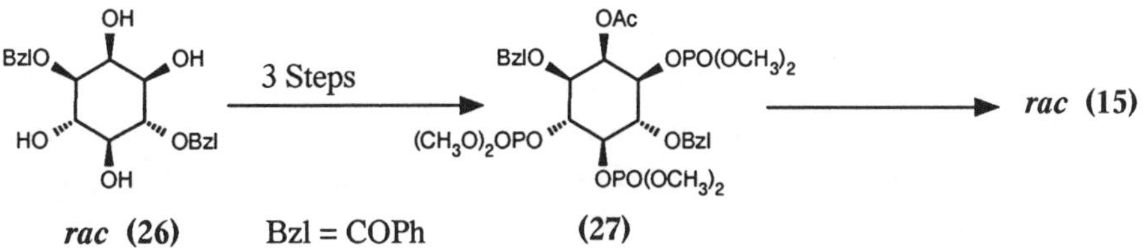

Fig 6 - 5

A very short synthesis of *rac*-1,4,5-IP$_3$ is possible, *Fig 6-5*, using the tetrol (26) obtained from the 1,2;4,5-bisisopropylidine acetal (11) by benzoylation and acetal cleavage with TFA, as starting material.[14] Phosphitylation of (26) with 3.3 equiv. of dimethyl chlorophosphate, followed by acylation of the crude reaction product, and oxidation of the phosphites to the phosphates with H$_2$O$_2$ gave the fully protected trisphosphate (27), in *ca* 94% yield and 95% purity. This selective phosphitylation avoids the need for a hydroxyl protecting group strategy. Deprotection with HBr in acetic acid, and ester hydrolysis provided *rac*-1,4,5-IP$_3$ (15) of *ca* 95% purity.

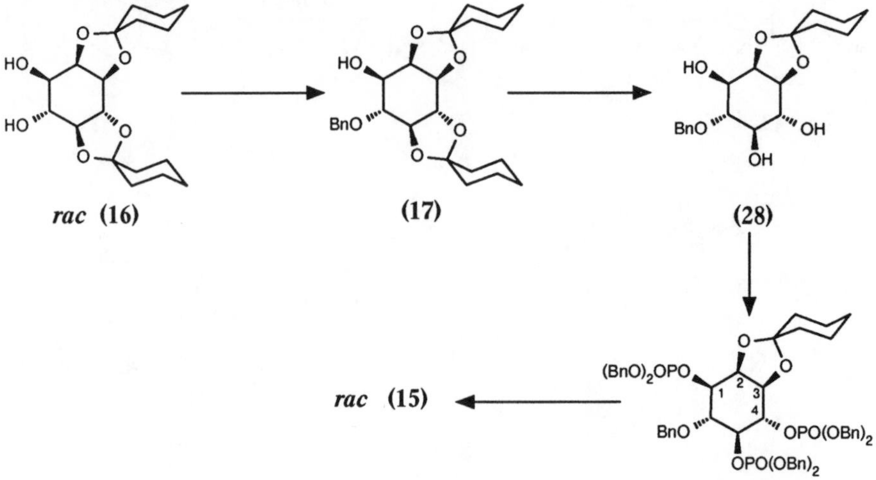

Fig 6 - 6

During the development of efficient phosphorylation strategies for polyols in the inositol series,[5] Frazer-Reid and Yu achieved 97% chemoselectivity in the conversion of the bisacetal (16) into (17) *via* the use of a tin-mediated benzylation, *Fig 6-6*.[15] Subsequent hydrolysis of the *trans* acetal gave the racemic triol (28), *ie* the racemic form of (+)-(21), see *Fig 6-3*. Phosphorylation of this triol using their two-step, one-pot protocol involving N,N-diisopropyl dibenzyl phosphoramidite, followed by oxidation with mCPBA, gave the fully protected racemic compound (29), in excellent yield. Although deprotection of (29) was not reported, presumably hydrogenolysis followed by acid treatment would give *rac*-1,4,5-IP$_3$, in analogy to the deprotection of the optically active material previously reported. Frazer-Reid and Yu also reported the phosphorylation of triol (13), *Fig 6-2*, prepared by the route of Gigg *et al*,[8] using the above reagent combination, and subsequent hydrogenolysis to *rac*-1,4,5-IP$_3$ (15).

Starting from the 1,2;4,5-bisacetal, van Boom *et al* prepared the protected diol (30) in two steps, *Fig 6-7*.[16] Selective benzylation of (30) gave (31) in modest yield, which on allylation of the remaining free hydroxyl group and acetal hydrolysis, gave the diol (32). Resolution of (32) was accomplished by conversion to the diastereomeric monomenthoxyacetates (33) and (34). Hydrolysis of the chiral auxiliary in the desired diastereomer (33), followed by protection of the resulting diol as the dibenzyl ether gave the fully protected chiral derivative (35). Cleavage of the three allyl groups from (35) then gave the previously reported optically active triol (+)-(8). This triol was phosphorylated by a protocol similar to that reported by Frazer-Reid and Yu,[5] but using the reagent combination N,N-diethyl dibenzyl phosphoramidite and *tert*-butyl hydroperoxide (*tert*-BuOOH), which van Boom's group had

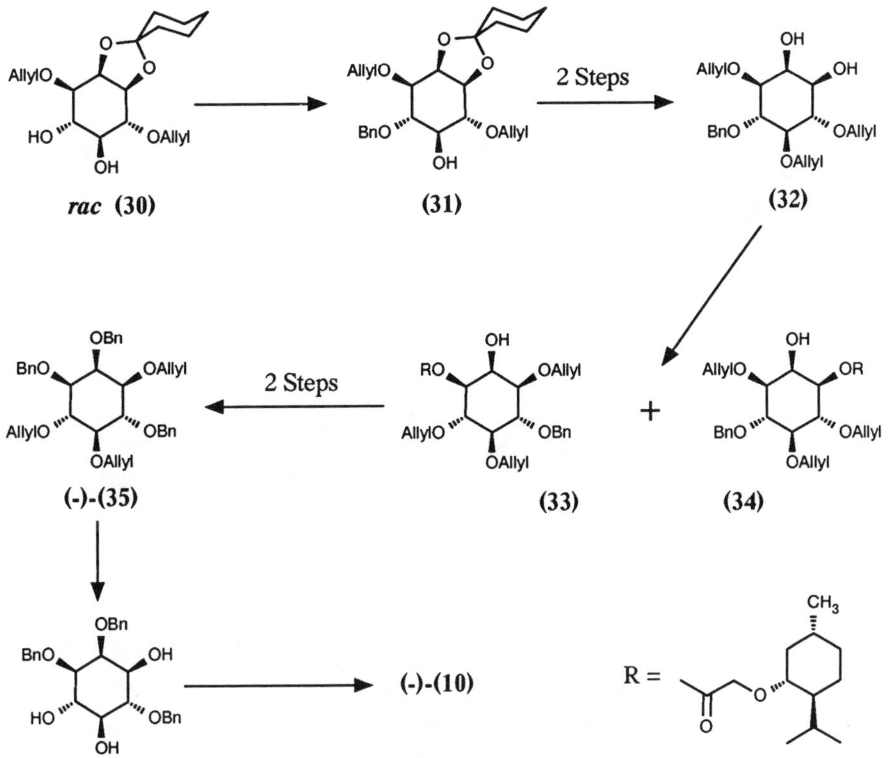

Fig 6 - 7

previously shown to be very effective in the preparation of phosphopeptides. The two groups of authors disagree over the relative merits of N,N-diethyl versus N,N-diisopropyl phosphoramidite, and the use of mCPBA rather than *tert*-BuOOH as oxidising agent in these reactions. Deprotection of the resulting fully benzylated derivative gave (-)-1,4,5-IP$_3$, (-)-(10), in excellent yield.

An unusual alternative strategy for the synthesis of the inositol phosphates involves phosphorylation at a very early stage in the synthetic sequence, *Fig 6-8*.[17] Ozaki *et al* phosphorylated the diol (36), obtained in three conventional steps from *myo*-inositol, using *n*-butyl lithium and tetrabenzyl pyrophosphate to obtain the fully protected bisphosphate (37). Hydrolysis of the cyclohexylidine acetal group then gave the racemic diol (38). Highly selective silylation (95%) of the more reactive equatorial hydroxyl group in (38) produced alcohol (39), which on benzoylation of the free hydroxyl group and cleavage of the silyl protection group gave alcohol (40). The final phosphate group was then introduced *via* sequential treatment with PCl$_3$ and BnOH to give the dibenzyl phosphite, which was oxidised to the phosphate level with *tert*-BuOOH. Deprotection of the resulting fully protected trisphosphate then gave *rac*-1,4,5,-IP$_3$ (15). The resolution of the intermediate diol (38) was

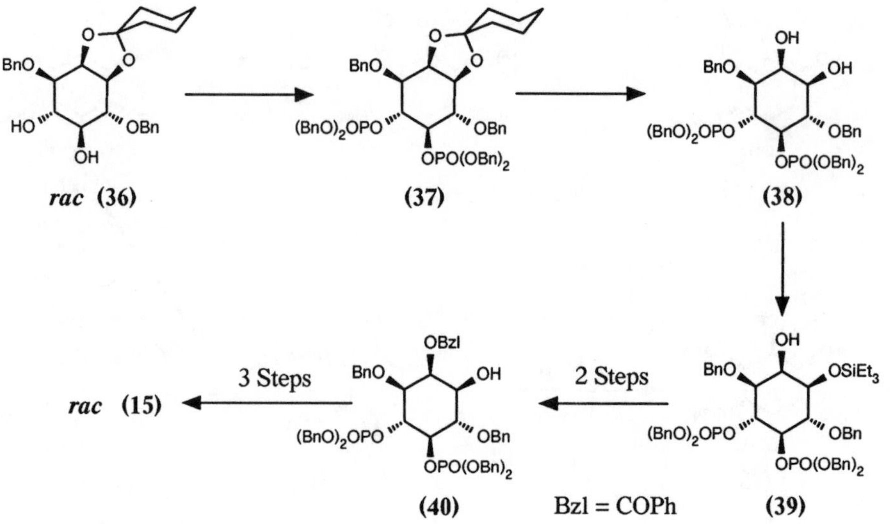

Fig 6 - 8

also reported *via* separation of the diastereomeric monomenthoxy acetates formed by selective reaction at the equatorial C-1 hydroxyl group.[17]

The fundamental approach to the *myo*-inositol skeleton adopted by Ley *et al* involves the establishment of all of the relative stereochemistry of the inositol system from benzene. This approach has been outlined in Chapter 3, and provides an interesting route to 1,4,5-IP$_3$.[18] Treatment of the *meso*-diol (40), obtained from microbial oxidation of benzene, with phosgene produced the cyclic carbonate (41), *Fig 6-9* . Epoxidation of the diene (41) with mCPBA gave the *alpha*-epoxide (42) as the major product (>4:1), which underwent regioselective ring opening on treatment with benzyl alcohol under acidic conditions to give the alcohol (43). Benzylation of the free hydroxyl group then gave the fully protected derivative. Stereoselective epoxidation of the remaining double bond was achieved by removal of the carbonate group to expose the diol function, followed by a hydroxyl-directed epoxidation using mCPBA at pH 8.8 to produce the *beta*-epoxide, and reprotection of the diol function as the acetonide to give (44). Less than 10% of the unwanted *alpha*-epoxide was formed in this sequence. As predicted by conformational analysis, regioselective epoxide opening of (44) with Ley's sterically unencumbered hydroxide equivalent DDEO gave the desired alcohol (45) in acceptable yield. The alcohol (45) has all of the relative stereochemistry of the inositol system in place, and also has all of the hydroxyl groups suitably differentiated for synthetic purposes. Thus, removal of the two benzyl ethers from (45), followed by phosphorylation of the resulting triol with *n*-butyl lithium / tetrabenzyl pyrophosphate, gave the fully protected trisphosphate (46). Deprotection by hydrogenolysis

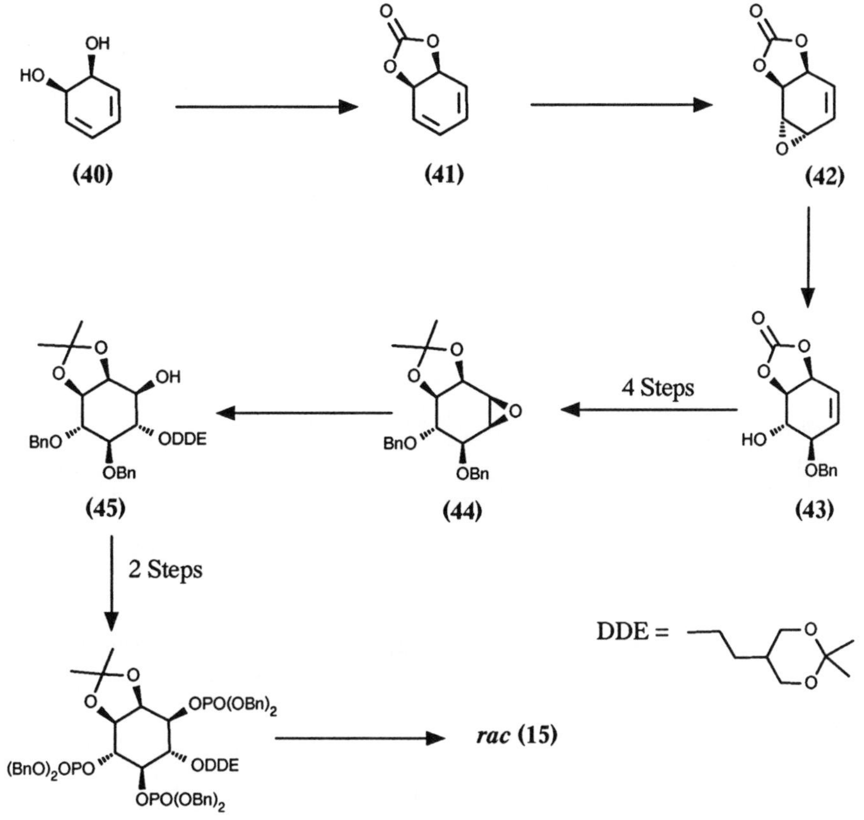

Fig 6 - 9

then provided *rac*-1,4,5-IP₃, (15).

The above synthetic approach has been further developed to provide the individual

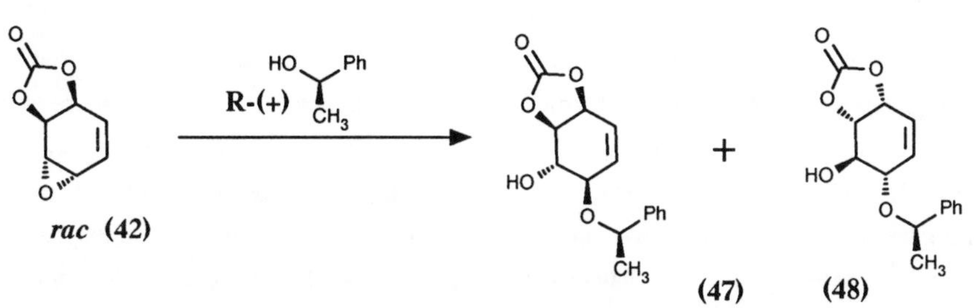

Fig 6 - 10

enantiomers of 1,4,5-IP₃.[18] The key to this chiral synthesis is the regioselective ring opening of the epoxide (42) with the homochiral alcohol (R)-(+)-*sec*-phenethyl alcohol, *Fig 6-10*, giving a mixture of the diastereomeric alcohols (47) and (48). On separation, these diastereomers represent equivalents of the enantiomers of alcohol (43), bearing a protecting group that happens to be chiral in place of the benzyl group. As the phenethyl protecting group is very similar to the benzyl group in stability, use of (47) and (48) as intermediates in place of (43) in the route shown in detail in *Fig 6-9* gives access to (-) and (+)-1,4,5-IP$_3$ respectively, *Fig 6-11*.

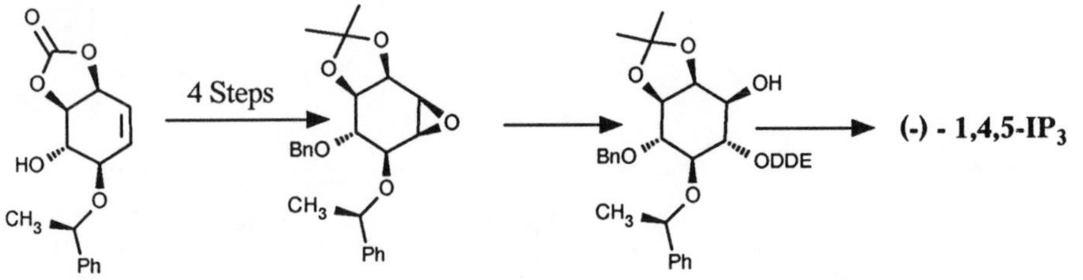

Fig 6 - 11

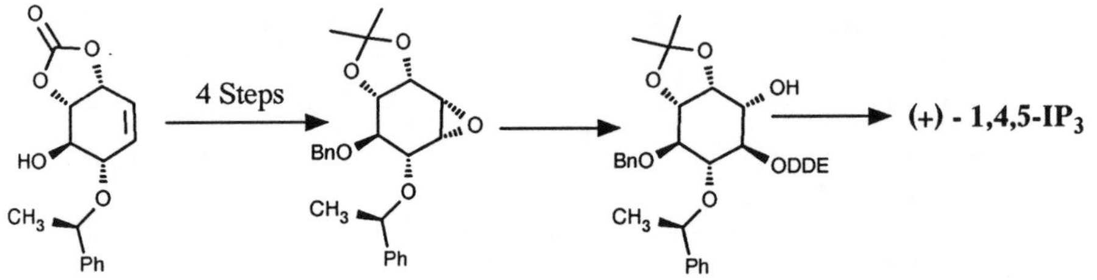

A more classical approach was used by Stepanov *et al*,[19] who followed an "early phosphorylation" strategy essentially identical to that outlined earlier in *Fig 6-8*. Phosphorylation of the racemic diol (49) [*cf* (36)] with dianilidophosphoryl chloride, followed by hydrolysis of the acetal group gave the diol (50), *Fig 6-12*. This diol was resolved by separation of the diastereomeric monomannose orthoesters (as pioneered by this Russian group, see Chapter 3), formed by preferential reaction at the more reactive equatorial hydroxyl group. Removal of the orthoester from the desired diastereoisomer then gave the desired enantiomer of (50) *ie* (+)-(51). Selective phosphorylation of (51) with dianilidophosphoryl chloride gave predominantly the desired equatorial phosphate (52). Sequential deprotection of this material provided (-)-1,4,5-IP$_3$, (-)-(10) in low yield, and requiring purification by HPLC. A similar sequence of reactions provided the enantiomeric material (+)-1,4,5-IP$_3$ from the other diastereomeric orthoester.

Cholesterol esterase shows interesting stereo- and regiospecificity in the hydrolysis of the diacetate (53) derived from the bisacetal (16),[20] allowing isolation of optically active diol (+)-(54), *Fig 6-13*. This material has been converted into (-)-1,4,5-IP$_3$, (-)-(10), by essentially

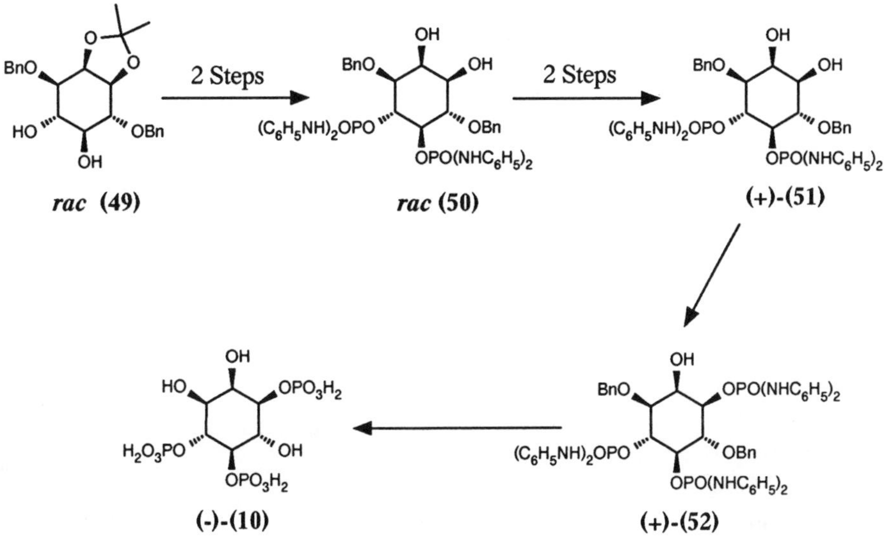

Fig 6 - 12

the same route as that described by Frazer-Reid and Yu, *Fig 6-6.*[5]

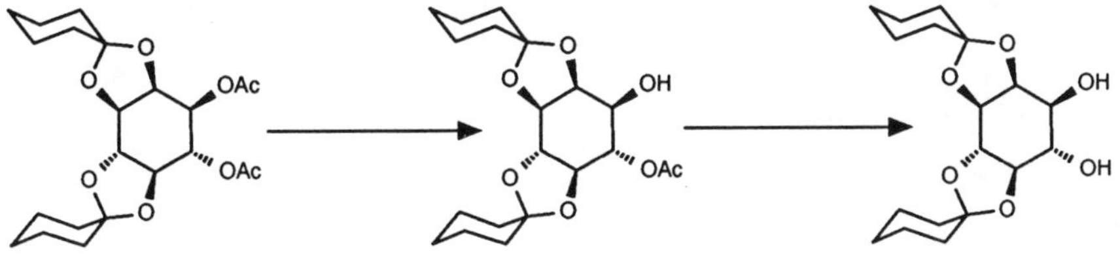

Fig 6 - 13

An alternative strategy in chiral synthesis involves the use of chiral starting materials. This approach was used by Falck and Yadagiri in their synthesis of (-)-1,4,5-IP$_3$ from the naturally occurring chiral starting material (-)-quinic acid.[21] Commercial (-)-quinic acid (55) was converted into the ester (56) *via* a known four-step procedure, *Fig 6-14*. Protection of the free hydroxyl function, followed by reduction of the ester to the primary alcohol, and selenylation then gave (57). Oxidation of (57) to the allylic selenoxide allowed rearrangement to the secondary alcohol, which was protected as its benzyl ether (58). Transformation of (58) into the silyl enol ether (59) was achieved by low temperature ozonolysis, followed by trapping with excess TBDMS chloride. Hydroboration of (59) occurred from the less hindered face, as expected, and peroxide oxidation produced the protected cyclitol (60). Removal of the silyl protecting groups gave the known triol (+)-(21), which on phosphorylation and deprotection by previously reported procedures[11] gave (-)-1,4,5-IP$_3$, (-)-(10).

A chiral synthesis of both enantiomers of 1,4,5-IP$_3$, starting from the much less readily

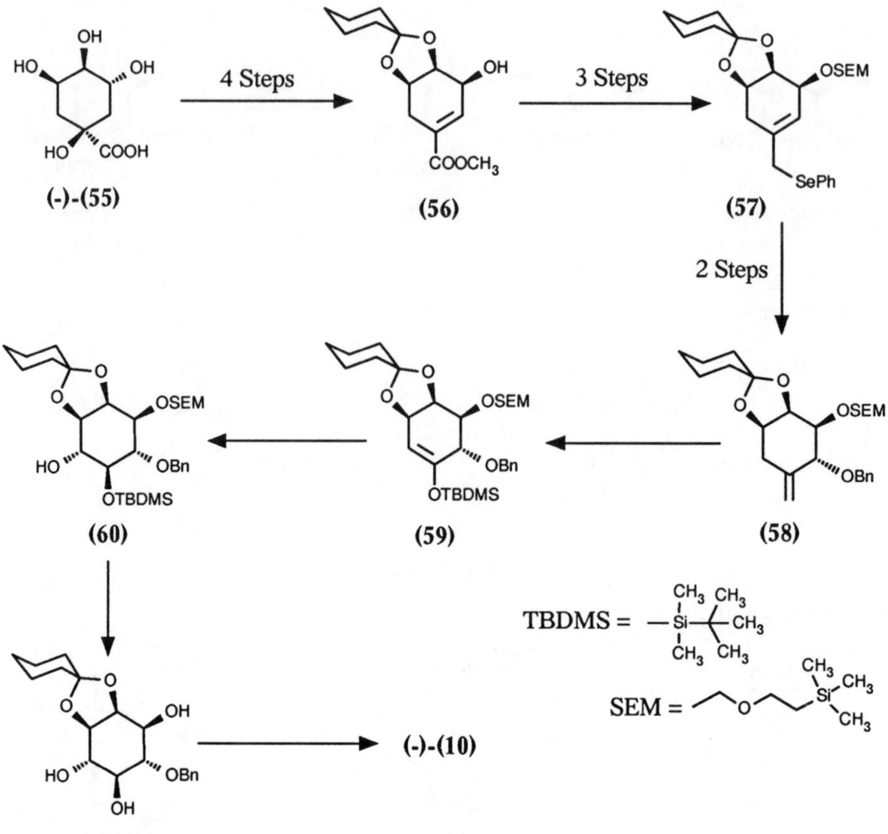

Fig 6 -14

available natural products D-pinitol and L-quebrachitol has been reported by Tegge and Ballou.[22] The starting point in this strategy is the conversion of D-pinitol into D-*chiro*-inositol, *Fig 6-15*, and L-quebrachitol into L-*chiro*-inositol by demethylation with HI. D-*chiro*-inositol was then converted into D-(-)-1,4,5-IP$_3$ by the route shown in *Fig 6-15*, whilst repetition of the synthetic sequence using the L-*chiro* isomer gave L-(+)-1,4,5-IP$_3$. Demethylation of D-pinitol (61) to D-*chiro*-inositol (62), followed by conversion to the bisacetal gave (63), the more stable *cis,cis* compound. Benzylation of the free hydroxyl groups, and hydrolysis of the acetals then gave the *chiro*-inositol dibenzyl ether (64). Benzoylation of (64) followed by preparative HPLC of the mixture of products gave the tribenzoyl alcohol (65) as the major product. Inversion of the free hydroxyl group was accomplished *via* conversion into the trifluoromethane sulphonate ester and displacement to give the protected *myo*-inositol derivative (66) as the major product. Hydrogenolysis of the benzyl ethers led to the trisbenzoyl derivative (67) which was cleanly phosphorylated using N,N-diisopropyl dibenzyl phosphoramidite, followed by oxidation with mCPBA giving (68).

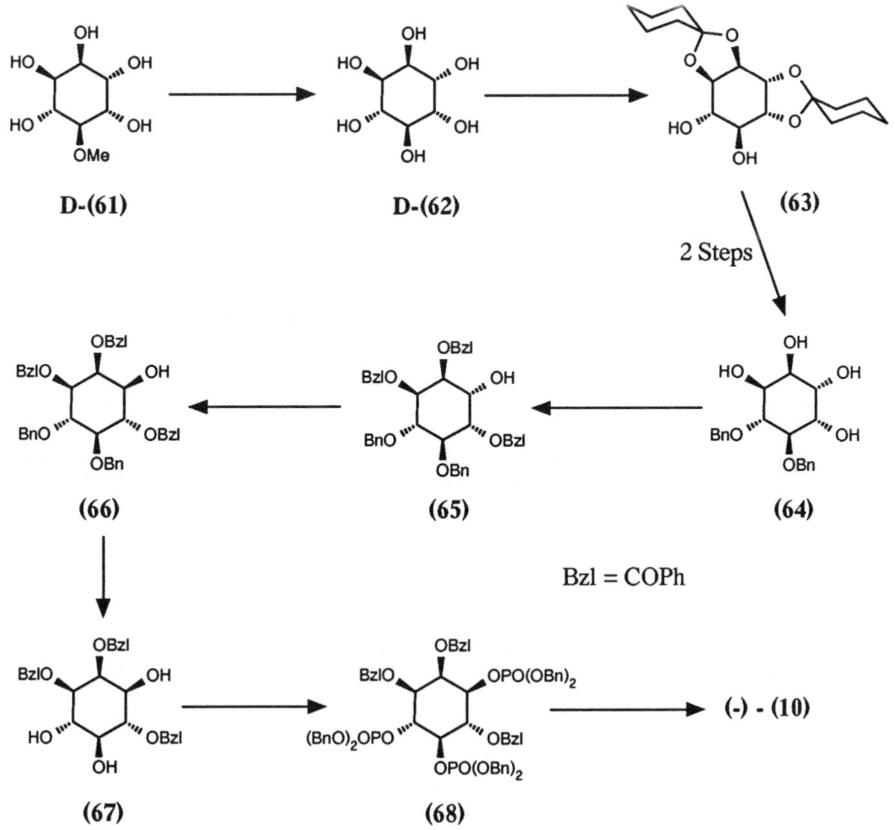

Fig 6 - 15

Sequential deprotection by hydrogenolysis followed by treatment with base then gave
(-)-1,4,5-IP$_3$, (-)-(10).

The synthetic approach of Carless and Busia *via* conduritol B derivatives (see chapter 5)
may also be applied to trisphosphate synthesis.[23] [4 + 2] Addition of singlet oxygen to the
protected *trans* diol (69) produces the protected tetrol (70), *Fig 6-16*. Oxidation of (70) to a
separable mixture of (71) and (72), followed by reduction of the desired isomer (72) with
NaBH$_4$/CeCl$_3$ then gave the diol (73). Benzylation of (73) followed by *cis* hydroxylation
using the catalytic OsO$_4$ system then produced the protected *myo*-inositol derivative (74).
Selective protection of the equatorial position in (74) gave (75) as the major product, which
on benzylation and removal of the MEM groups gave the known triol (13). Phosphorylation
and deprotection of (13) by known methods then produced *rac*-1,4,5-IP$_3$, (15).

Ozaki *et al* have reported a highly selective esterification method for resolution of
1,3,5-tri-O-benzoyl *myo*-inositol,[24] and have applied this to the synthesis of (-)-1,4,5,-IP$_3$, *Fig*
6-17.[25] The starting inositol derivative 1,3,5-tri-O-benzoyl *myo*-inositol (76) is obtained in

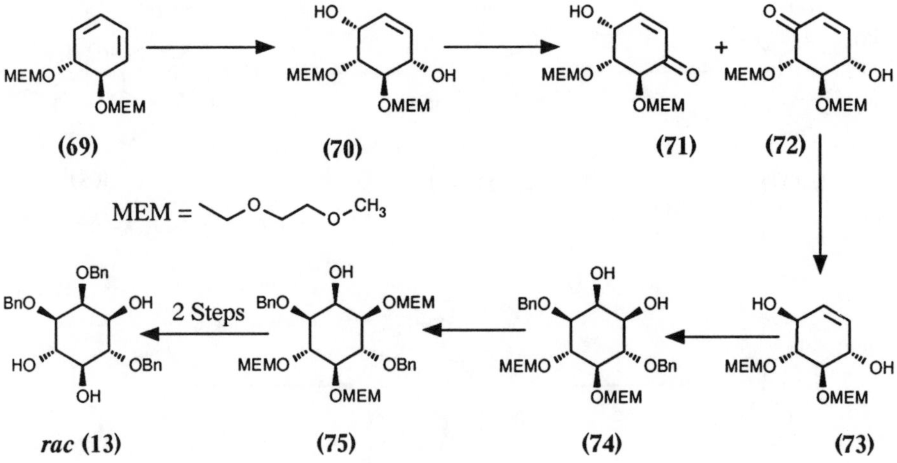

(69)   **(70)**   **(71)**   **(72)**

MEM =

rac (13)   **(75)**   **(74)**   **(73)**

Fig 6 - 16

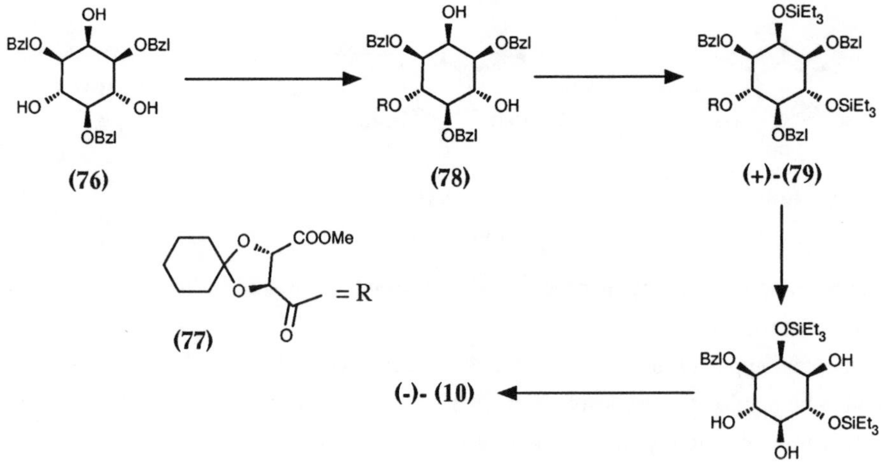

(76)   **(78)**   **(+)-(79)**

(77)

(-)- (10)   **(80)**

Fig 6 - 17

low yield (15%) by direct benzoylation of *myo*-inositol. Reaction with the specific tartrate derivative shown (77) proceeds with remarkable enantioselectivity to give (78) in 57% yield and 96% diastereomeric excess. Silylation of the two free hydroxyl groups then gave the disilyl ether (79), which was optically pure after one recrystallisation. Highly selective cleavage of the acyl groups using methylmagnesium bromide in ether left the 3-benzoyl group intact to give (80) in high yield. Phosphorylation of (80) with the new reagent O-xylylene phosphoramidite, followed by deprotection then gave (-)-1,4,5-IP$_3$, (-)-(10).[25]

6.3 Inositol 1,3,4-Trisphosphate

A number of syntheses of inositol 1,3,4-trisphosphate (1,3,4-IP$_3$) have been reported using essentially the same strategies as the syntheses of the 1,4,5- isomer above. In particular, several syntheses have hinged on the preparation of the racemic triol (84).

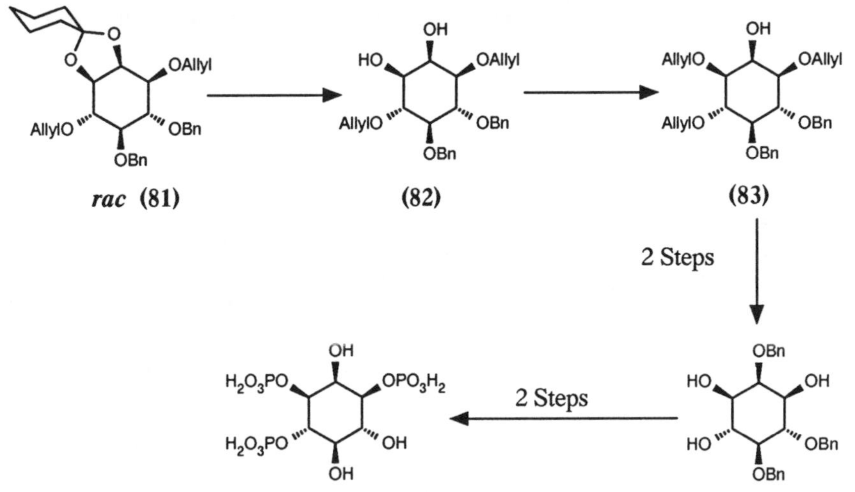

Fig 6 - 18

Bis-allylation of the 1,2;4,5-biscyclohexylidine acetal, followed by selective hydrolysis of the *trans* acetal, and benzylation of the resulting diol gives the fully protected inositol (81), *Fig 6-18*. This approach is directly analogous to that shown in *Fig 6-1* for 1,4,5-IP$_3$. Hydrolysis of the *cis* acetal then gives diol (82) which may be selectively allylated at the more reactive equatorial hydroxyl group to produce alcohol (83). Benzylation of (83), followed by cleavage of the allyl groups then gives the key triol *rac*-(84). This route has been followed starting from both the biscyclohexylidine acetal,[5,11,26] and the analogous bisisopropylidine acetal.[8] Phosphorylation of the alkoxide derived from (84) using tetrabenzylpyrophosphate, followed by hydrogenolysis gave *rac*-1,3,4-IP$_3$, (85) in high yield.[11,26] Other successful phosphorylation strategies for this triol include the use of the two-step, one-pot protocol involving N,N-diisopropyl dibenzyl phosphoramidite followed by oxidation with mCPBA,[5] and phosphitylation with ClP(OCH$_2$CH$_2$CN)$_2$ followed by oxidation.[27]

An interesting convergent approach involves the use of the two bisacetals (16) and (86).[28] By reversing the order of introduction of the benzyl and *p*-methoxybenzyl protecting groups, both (16) and (86) may be converted into the racemic intermediate (89), *Fig 6-19*, *via* protection, giving (87) and (88), selective hydrolysis of the *trans* acetal, and protection of the resulting diol with the second protecting group. Cleavage of the remaining acetal gives the racemic diol, which may be resolved *via* its monomenthoxy acetates [*cf* resolution of (3) in *Fig 6-1*] to give the desired enantiomer (-)-(90). Selective incorporation of the MOM group at the equatorial position, followed by benzylation gave the fully protected enantiomerically pure derivative (91). Removal of the *p*-methoxybenzyl and MOM groups then gave the desired triol, which on phosphorylation using tetrabenzylpyrophosphate and deprotection gave (-)-1,3,4-IP$_3$, (-)-(92).

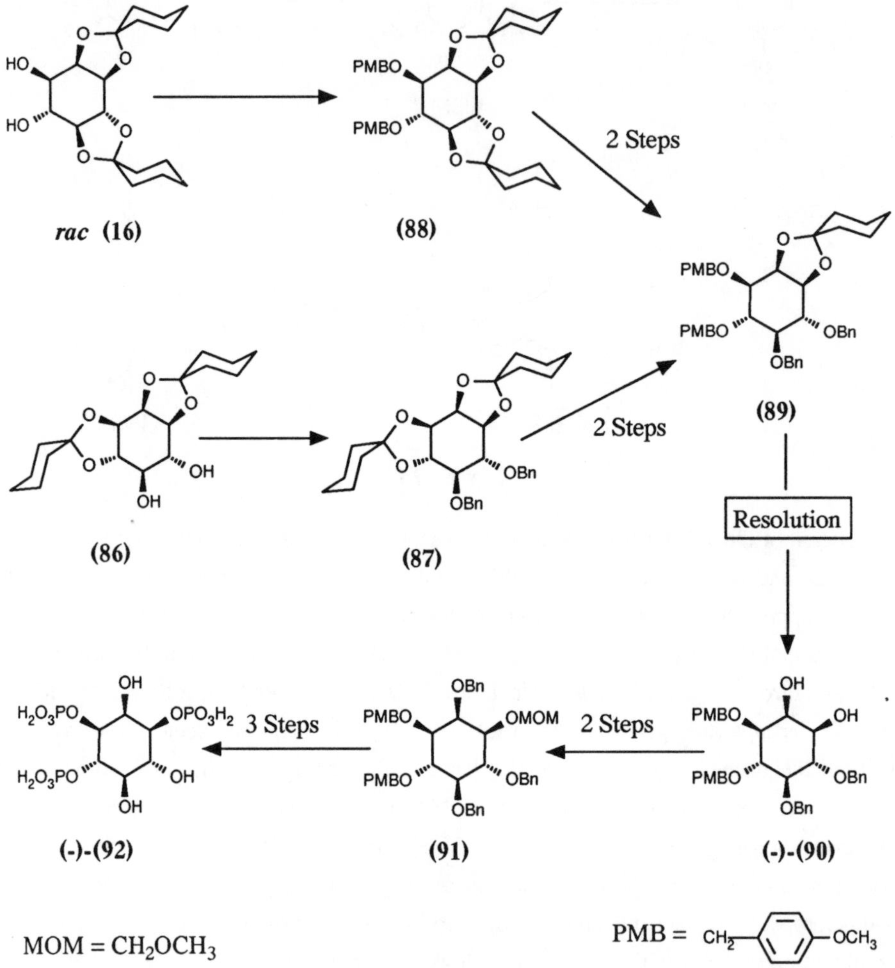

MOM = CH$_2$OCH$_3$

PMB = CH$_2$—⟨ ⟩—OCH$_3$

Fig 6 - 19

Selective protection of 1,2-O-cyclohexylidene-*myo*-inositol with 1,3-dichloro-1,1,3,3-tetraisopropyldisiloxane (TIPS; see Chapter 4), followed by benzoylation of the resulting

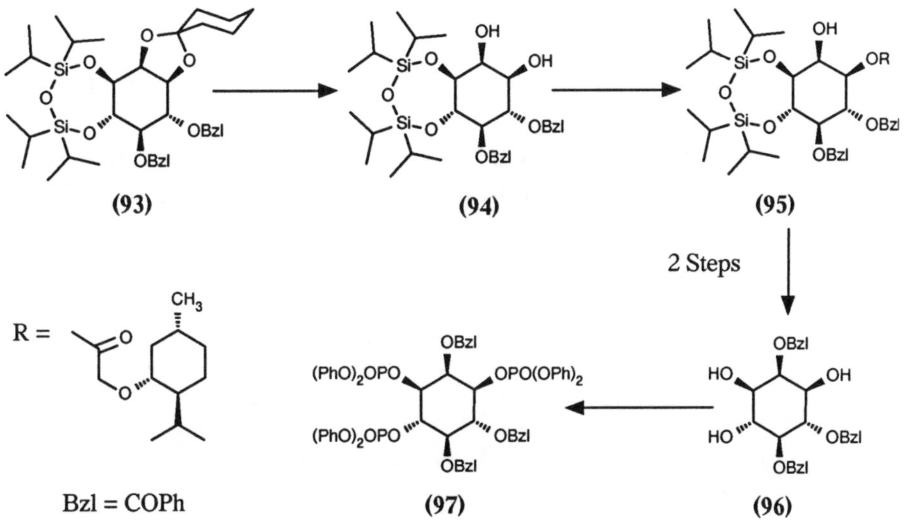

Fig 6 - 20

diol gives the fully protected inositol (93), *Fig 6-20*.[29] Hydrolysis of the acetal produces the diol (94), which reacts selectively at the equatorial position to form the diastereomeric monomenthoxy acetate derivatives. Benzoylation of the desired isomer (95), followed by cleavage of the TIPS and menthoxy acetate groups then provides the key optically active triol (96). This triol was phosphorylated using diphenyl phosphorochloridite coupled with oxidation to the phosphate level to give (97). The deprotection of (97) to give optically active 1,3,4-IP$_3$ has not been reported at this time.

6.4 Inositol 2,4,5-Trisphosphate

Given the above methodology, the synthesis of inositol 2,4,5-trisphosphate (2,4,5-IP$_3$) is relatively straightforward.[11,26] The preparation of diol (3) has been outlined in *Fig 6-2*. Selective benzylation of (3) at the equatorial position, followed by cleavage of the allyl groups gives triol (98) directly, *Fig 6-21*. Phosphorylation of the anion of (98) with

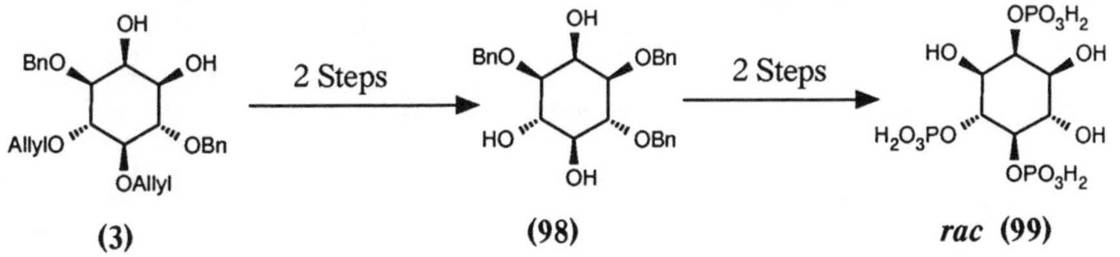

Fig 6 - 21

tetrabenzylpyrophosphate and deprotection by hydrogenolysis gave *rac*-2,4,5-IP$_3$, (99).

A formal synthesis of the enantiomers of 2,4,5-IP$_3$ has been reported involving resolution of diol (3) *via* its monomenthoxy acetates (see *Fig 6-1*).[30] This synthesis was not successfully concluded due to problems with phosphorylation technology.

Direct phosphorylation of the alcohol (39), whose synthesis is described in *Fig 6-8, via* sequential treatment with PCl$_3$ and BnOH to give the dibenzyl phosphite, and oxidation to the phosphate level with *tert*-BuOOH gives the fully protected phosphate (100), *Fig 6-22*.[17] Deprotection by hydrogenolysis leads to *rac*-2,4,5-IP$_3$, (99).

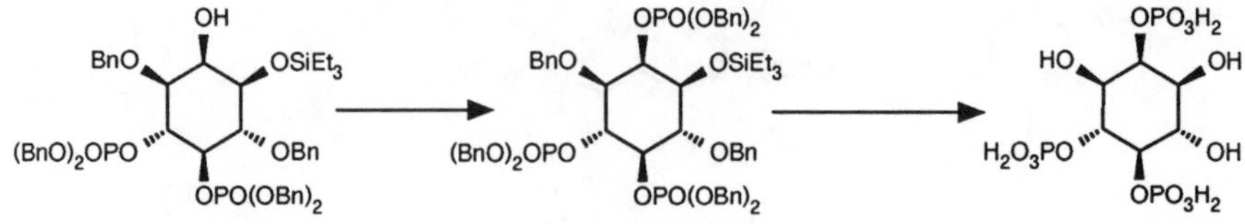

Fig 6 - 22

Carless and Busia have reported the isolation of alcohol (101) in *ca* 10% recovery during incorporation of a MEM group into diol (74), together with *ca* 60% of the desired isomer (75), *Fig 6-23*.[23] The synthesis of 1,4,5-IP$_3$ from (75) is shown in *Fig 6-16*. Benzylation of (101), followed by cleavage of the MEM groups provides the triol (98). Phosphorylation of the anion of (98) using tetrabenzylpyrophosphate, and deprotection by hydrogenolysis then gives *rac*-2,4,5-IP$_3$ (99).

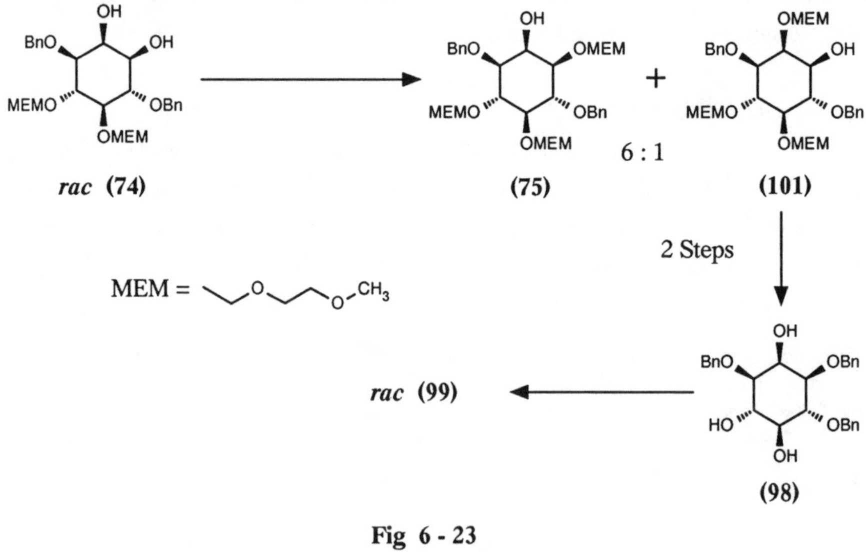

Fig 6 - 23

References

1) S.Ozaki, Y.Watanabe, T.Ogasawara, Y.Kondo, N.Shiotani, H.Nishii and T.Matsuki, *Tetrahedron Lett.,* 1986, **27**, 3157.

2) R.Gigg and C.D.Warren, *J.Chem.Soc.(C),* 1969, 2367.

3) D.C.Billington and R.Baker, *J.Chem.Soc.Chem.Commun.,* 1987, 1011.

4) Y.Watanabe, H.Nakahira, M.Bunya and S.Ozaki, *Tetrahedron Lett.,* 1987, **28**, 4179.

5) K.-L. Yu and B.Frazer-Reid, *Tetrahedron Lett.,* 1988, **29**, 979.

6) J.Gigg, R.Gigg, S.Payne and R.Conant, *Carbohydrate Res.,* 1985, **140**, c1-c3.

7) A.M.Cooke, B.V.L.Potter and R.Gigg, *Tetrahedron Lett.,* 1987, **28**, 2305; A.M.Cooke, B.V.L.Potter and R.Gigg, *Biochem.Soc.Trans.,* 1987, **15**, 904.

8) J.Gigg, R.Gigg, S.Payne and R.Conant, *J.Chem.Soc.Perkin.Trans.1,* 1987, 423.

9) S.David and S.Hanessian, *Tetrahedron,* 1985, **41**, 643.

10) J.P.Vacca, S.J.de Solms and J.R.Huff, *J.Am.Chem.Soc.,* 1987, **109**, 3478.

11) J.P.Vacca, S.J.de Solms, J.R.Huff, D.C.Billington, R.Baker, J.J.Kulagowski and I.M.Mawer, *Tetrahedron,* 1989, **45**, 5679.

12) C.B.Reese and J.G.Ward, *Tetrahedron Lett.,* 1987, **28**, 2309.

13) A.E.Stepanov, B.A.Klyashchitskii, V.I.Shvets and R.P.Evstigneeva, *Bioorg.Khim.,* 1976, **2**, 1627.

14) J.L.Meek, F.Davidson and F.W.Hobbs, *J.Am.Chem.Soc.,* 1988, **110**, 2317.

15) N.Nagashima and N.Ohno, *Chem.Lett.,* 1987, 141.

16) C.E.Dreef, R.J.Tuinman, C.J.J.Elie, G.A.van der Marel and J.H.van Boom, *Recl.Trav.Chim.Pays-Bas,* 1988, **107**, 395.

17) Y.Watanabe, T.Ogasawara, H.Nakahira, T.Matsuki and S.Ozaki, *Tetrahedron Lett.,* 1988, **29**, 5259.

18) S.V.Ley and F.Sternfeld, *Tetrahedron Lett.,* 1988, **29**, 5305; S.V.Ley, M.Parra, A.J.Redgrave and F.Sternfeld, *Tetrahedron,* 1990, **46**, 4995.

19) A.E.Stepanov, O.B.Runova, G.Schlewer, B.Spiess and V.I.Shvets, *Tetrahedron Lett.,* 1989, **30**, 5125.

20) Y-C.Liu and C-S.Chen, *Tetrahedron Lett.,* 1989, **30**, 1617.

21) J.R.Falck and P.Yadagiri, *J.Org.Chem.,* 1989, **54**, 5851.

22) W.Tegge and C.E.Ballou, *Proc.Natl.Acad.Sci.USA,* 1989, **86**, 94.

23) H.A.J.Carless and K.Busia, *Tetrahedron Lett.,* 1990, **31**, 3449.

24) Y.Watanabe, A.Oka, Y.Shimizu and S.Ozaki, *Tetrahedron Lett.,* 1990, **31**, 2613.

25) Y.Watanabe, T.Fujimoto, T.Shinohara and S.Ozaki, *J.Chem.Soc.Chem.Commun.,* 1991, 428.

26) S.J.de Solms, J.P.Vacca and J.R.Huff, *Tetrahedron Lett.,* 1987, **28**, 4503.

27)   C.E.Dreef, G.A.van der Marel and J.H.van Boom, *Recl.Trav.Chim.Pays-Bas,* 1987, **106**, 161.

28)   S.Ozaki, M.Kohno, H.Nakahira, M.Bunya and Y.Watanabe, *Chem.Lett.*, 1988, 77.

29)   Y.Watanabe, M.Mitani, T.Morita and S.Ozaki, *J.Chem.Soc.Chem.Commun.*, 1989, 482.

30)   Y.Watanabe, T.Ogasawara, N.Shiotani and S.Ozaki, *Tetrahedron Lett.*, 1987, **28**, 2607.

Chapter 7

Synthesis of Inositol Tetrakisphosphates

7.1 Introduction

The discovery of an alternative pathway for the metabolism of (-)-1,4,5-IP$_3$ *via* phosphorylation at the 3- position giving (-)-inositol 1,3,4,5-tetrakisphosphate (1,3,4,5-IP$_4$), see Chapter 2, followed by sequential dephosphorylation leading to free inositol, resulted in active interest in the synthesis of this tetrakisphosphate and its isomers. The syntheses of these molecules will be presented in chronological order, arranged by substitution pattern, as for the trisphosphates in Chapter 6. Most of the synthetic effort has focused on the naturally occurring metabolite (-)-inositol 1,3,4,5-tetrakisphosphate, and in general the other isomers which have been prepared have resulted from the phosphorylation of available synthetic intermediates, rather than representing synthetic targets themselves.

7.2 Inositol 1,3,4,5-Tetrakisphosphate

The first reported synthesis of 1,3,4,5-IP$_4$ [1,2] took advantage of the highly specific chelation-controlled alkylation reactions of inositol orthoformate (1), *Fig 7-1*, previously discussed in Chapter 3. Selective mono-allylation of (1) with NaH and allyl bromide in DMF gave the 4-

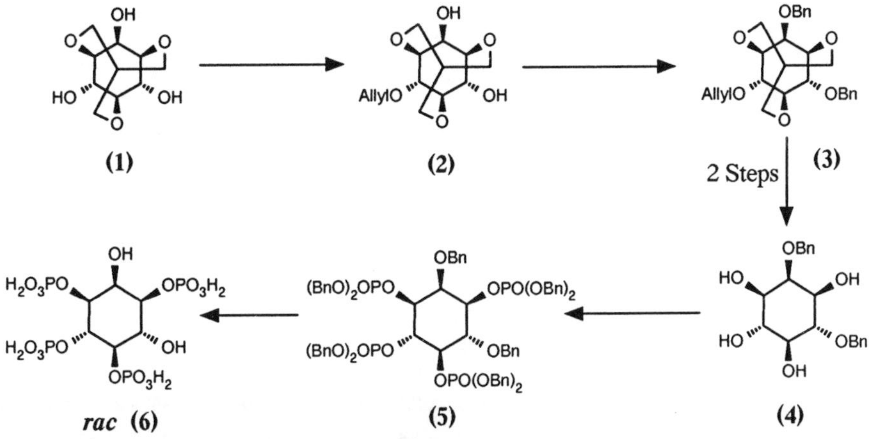

Fig 7 - 1

allyl derivative (2) in high yield, which was perbenzylated to give (3). Isomerisation of the allyl group to the enol ether using Wilkinson's catalyst,[3] followed by acid hydrolysis of both the enol ether and orthoformate protecting groups gave the racemic tetrol (4) directly. Phosphorylation of (4) using tetrabenzylpyrophosphate/ NaH/ imidazole produced the fully protected tetrakisphosphate (5) in high yield, despite the presence of three adjacent phosphorylation sites. Deprotection of (5) by hydrogenolysis then gave *rac*-1,3,4,5-IP$_4$ in quantitative yield. An essentially identical synthesis of *rac*-1,3,4,5-IP$_4$ which uses a benzyloxymethyl ether in place of an allyl group for temporary protection of the 4- position has been reported.[4] The tetrol (4) may also be efficiently phosphitylated using N,N-diisopropyl dibenzyl phosphoramidite,[5] followed by oxidation of the tetrakisphosphite to the tetrakisphosphate, and finally deprotected to give *rac*-1,3,4,5-IP$_4$.

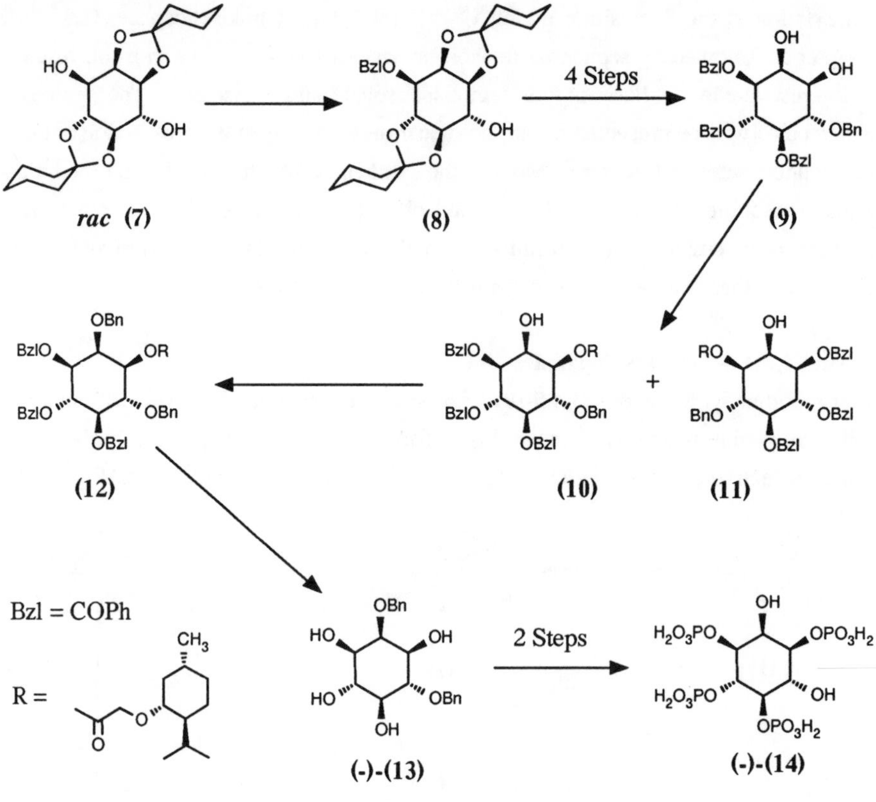

Fig 7 - 2

A more classical approach [6] uses the bisacetal (7), see Chapter 3, as starting material. Selective benzoylation of (7) using benzoyl imidazole/ CsF gave the alcohol (8), *Fig 7-2*. Benzylation of the free hydroxyl group using benzyltrichloroacetimidate and a catalyst, followed by selective hydrolysis of the less stable *trans* acetal, benzoylation of the resulting diol, and subsequent hydrolysis of the remaining *cis* acetal gave the racemic diol (9). Conversion of (9) into its diastereomeric monomenthoxy acetates by selective reaction at the equatorial 1- hydroxyl group, followed by separation of the diastereomers (10) and (11) gave optically pure intermediates. The desired diastereomer (10) was then benzylated to give the fully protected inositol (12), which on hydrolysis of the four ester groups gave the optically pure tetrol (-)-(13). Phosphorylation of (-)-(13) using tetrabenzylpyrophosphate, followed by deprotection using hydrogenolysis gave (-)-1,3,4,5-IP$_4$, (-)-(14), identical to the natural material.

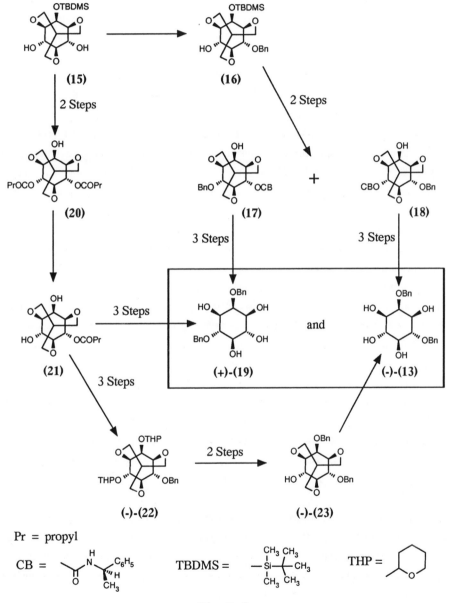

Fig 7 - 3

An elegant approach involves the preparation of the enantiomers of tetrol (4), *ie* (-)-(13) and (+)-(19), from the monosilyl orthoformate (15),[7] prepared from the orthoformate (1) in 60 % yield by the method originally outlined by Lee and Kishi.[8] Monobenzylation of (15) gives the racemic alcohol (16), which may be resolved *via* conversion into the diastereomeric carbamates (17) and (18), *Fig 7-3*. Separation of these diastereomers was only possible after desilylation and benzylation of the free hydroxyl group, and subsequent hydrolysis of the carbamate and orthoformate groups gave the enantiomeric tetrols (-)-(13) and (+)-(19) in optically pure form.

The same authors described an enzymic approach to the preparation of (-)-(13) and (+)-(19), *via* diesterification of (15) and subsequent desilylation to give the alcohol (20). Treatment of the alcohol (20) with pig liver esterase gave a high yield of the optically active diol (-)-(21) in 95% enantiomeric excess. Dibenzylation of (-)-(21), followed by ester and orthoformate hydrolysis then gave the optically active tetrol (+)-(19) directly. By taking advantage of the symmetry of the *myo*-inositol molecule, diol (-)-(21) may also be converted into the enantiomeric tetrol (-)-(13), *Fig 7-3*. Protection of diol (-)-(21) as its bistetrahydropyranyl (THP) ether, ester cleavage, and subsequent benzylation of the free hydroxyl group gave the fully protected inositol (-)-(22). Hydrolysis of the THP ethers, followed by selective benzylation at the less hindered equatorial hydroxyl group gave (-)-(23), which gave (-)-(13) on orthoformate cleavage. Phosphitylation,[5] oxidation and finally deprotection of these enantiomeric tetrols then gave the enantiomers of $1,3,4,5-IP_4$.

Starting from the 1,2;4,5-biscyclohexylidineacetal, van Boom *et al* prepared the protected diol (24) in two steps, *Fig 7-4*.[9] Selective benzylation of (24) gave (25) in modest yield, which on allylation of the remaining free hydroxyl group and acetal hydrolysis, gave the diol (26). Resolution of (26) was accomplished by conversion to the diastereomeric monomenthoxyacetates (27) and (28). Hydrolysis of the chiral auxiliary in the desired diastereomer (27), followed by selective allylation at the equatorial position *via* a cyclic tin intermediate gave the alcohol (+)-(29). Benzylation of the free hydroxyl group in (+)-(29), followed by removal of the allyl groups gave the optically active tetrol (-)-(13). Phosphitylation of (-)-(13) using N,N-diisopropyl dibenzyl phosphoramidite,[5] followed by oxidation of the tetrakisphosphite to the tetrakisphosphate, and finally deprotection gave (-)-$1,3,4,5-IP_4$, (-)-(14).

A somewhat less selective approach [10] involves the migration of a benzoyl group from the equatorial position of tetrol (30) to the neighbouring axial position to give (31), in modest yield, *Fig 7-5*. Phosphitylation of (31) with $ClP(OMe)_2$, followed by oxidation gave the protected tetrakisphosphate (32). Acidic hydrolysis of the phosphate esters, followed by basic hydrolysis of the benzoate esters then gave *rac*-$1,3,4,5-IP_4$, (6).

Ozaki *et al* have studied the direct benzoylation of *myo*-inositol, which under controlled

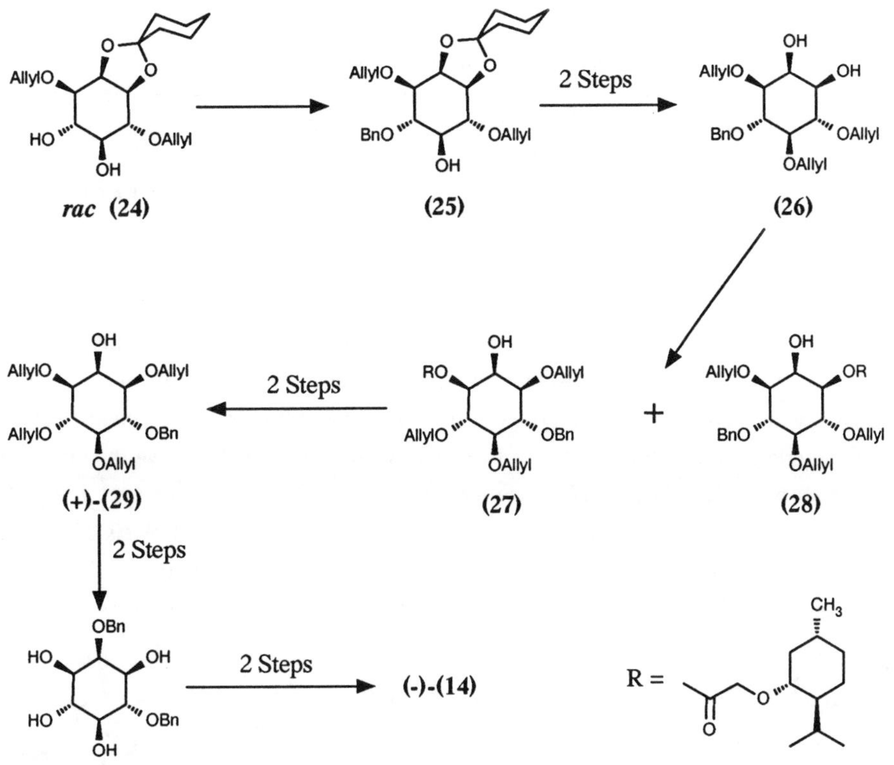

Fig 7 - 4

conditions can provide the 1,3,4,5-tetrabenzoate (33), *Fig 7-6*, in 33% isolable yield, see Chapter 3.[11] Benzylation of the two free hydroxyl groups of (33) using benzyl trichloroacetimidate in the presence of trifluoromethanesulphonic acid then provides the fully protected intermediate (34). Hydrolysis of the four ester groups in (34) then produces the

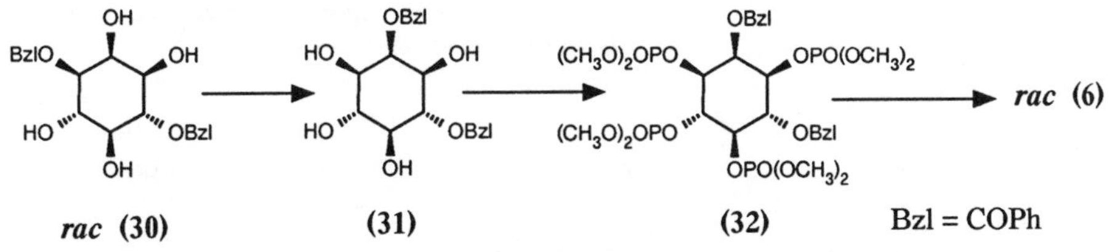

Fig 7 - 5

racemic tetrol (4), which on phosphitylation using an O-xylylene phosphoramidite and subsequent oxidation and deprotection gave *rac*-1,3,4,5-IP$_4$, (6).

Ozaki *et al* have also reported a highly selective esterification method for the resolution of 1,3,5-tri-O-benzoyl *myo*-inositol,[12] and have applied this to the synthesis of (-)-1,3,4,5,-IP$_4$,

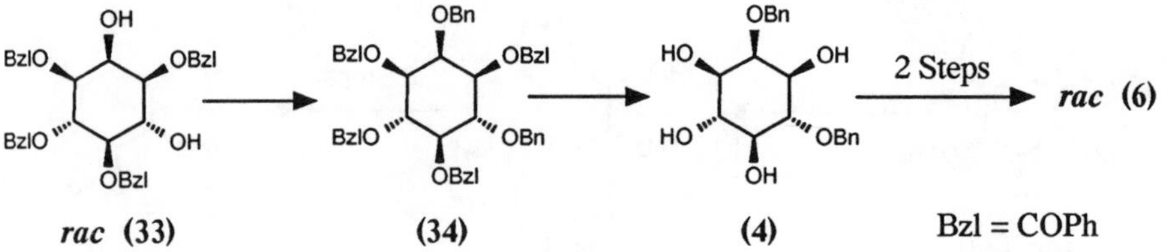

Fig 7 - 6

Fig 7-7 (and also to the synthesis of (-)-1,4,5-IP$_3$, see Chapter 6).[13] The starting inositol derivative 1,3,5-tri-O-benzoyl *myo*-inositol (35) is obtained in low yield (15%) by direct benzoylation of *myo*-inositol. Reaction with the specific tartrate derivative shown (39) proceeds with remarkable enantioselectivity to give (36) in 57% yield and 96% diastereomeric excess. Silylation of the two free hydroxyl groups then gave the disilyl ether (37), which was optically pure after one recrystallisation. Removal of the acyl groups from (37) was complicated by the attack of alkoxide, liberated during the hydrolysis of the esters, on the remaining silyl groups. Treatment of (37) with *ca* 35 equiv. ethylmagnesium bromide in diethyl ether under reflux however gave 89% recovery of optically active tetrol (38).

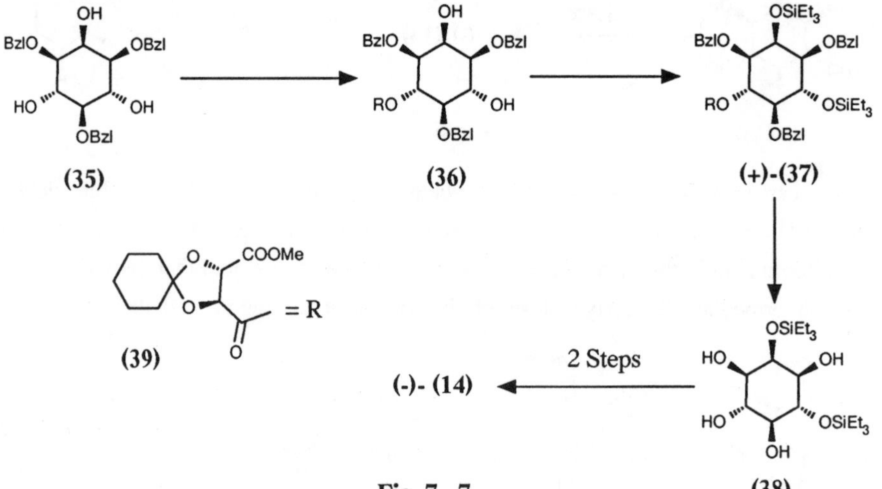

Fig 7 - 7

Phosphitylation with an O-xylylene phosphoramidite followed by oxidation to the phosphate level, and subsequent deprotection gave (-)-1,3,4,5-IP$_4$, (-)-(14).

7.3 Inositol 1,2,4,5-Tetrakisphosphate

Carless and Busia have used the intermediate diol (40) prepared during their synthesis of 1,4,5-IP$_3$ (see Chapter 6, *Fig 6-16*) to prepare racemic 1,2,4,5-IP$_4$.[14] Removal of the MEM protecting groups from (40) gave the racemic tetrol (41), *Fig 7-8*, which on phosphorylation with tetrabenzylpyrophosphate/ NaH/ THF and deprotection by hydrogenolysis gave *rac*-1,2,4,5-IP$_4$, (42). Meek *et al* had previously reported the synthesis of *rac*-1,2,4,5-IP$_4$ isolated as an impure product following phosphitylation, oxidation, and deprotection of 3,6-dibenzyl inositol.[10]

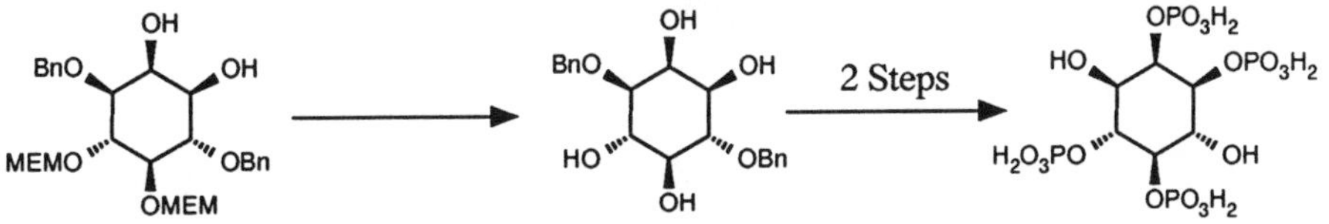

Fig 7 - 8

7.4 Inositol 1,3,4,6-Tetrakisphosphate

Ozaki *et al* obtained the symmetrical bis(disiloxane) derivative (43), *Fig 7-9*, directly in 66% yield by treatment of *myo*-inositol with 1,3-dichloro-1,1,3,3-tetraisopropyldisiloxane.[15] No other inositol derivatives were obtained from the reaction mixture. Benzoylation of (43) gave the 2,5-dibenzoate (44), which on treatment with HF in wet acetonitrile gave the tetrol (45). Phosphitylation of (45), followed by oxidation to the phosphate and deprotection then produced the *meso*-tetrakisphosphate 1,3,4,6-IP$_4$, (46), in high yield.

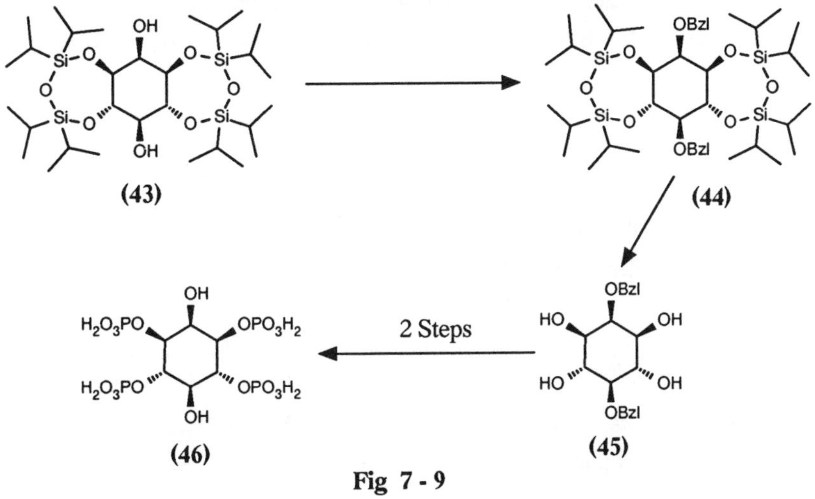

Fig 7 - 9

7.5 Inositol 1,4,5,6-Tetrakisphosphate

By direct phosphitylation of the well-known 1,2-cyclohexylidine acetal (47) with

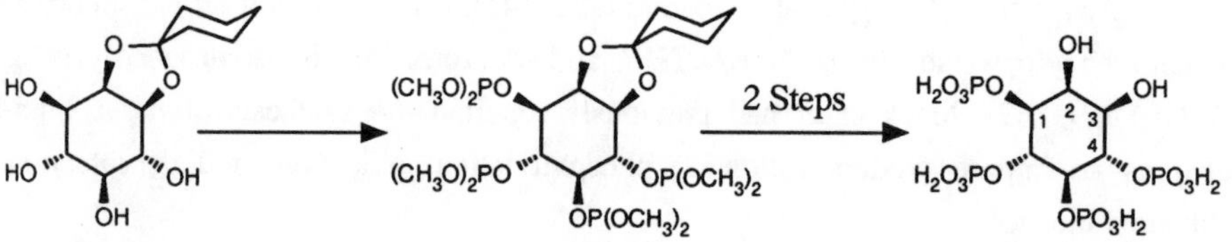

Fig 7 - 10

$(CH_3O)_2PCl$ [10] Meek *et al* obtained the protected phosphite (48), *Fig 7-10*. Oxidation to the protected phosphate, followed by deprotection of the methyl esters with bromotrimethylsilane, and subsequent self-catalysed hydrolysis of the acetal group produced *rac*-1,4,5,6-IP_4, (49).

References

1)   D.C.Billington and R.Baker, *J.Chem.Soc.Chem.Commun.,* 1987, 1011.

2)   D.C.Billington, R.Baker, J.J.Kulagowski, I.M.Mawer, S.J.de Solms, J.P.Vacca and J.R.Huff, *J.Chem.Soc.Perkin Trans. 1*, 1989, 1423.

3)   E.J.Corey and J.W.Suggs, *J.Org.Chem.,* 1976, **38**, 3224.

4)   S.J.de Solms, J.P.Vacca and J.R.Huff, *Tetrahedron Lett.,* 1987, **28**, 4503.

5)   K.-L. Yu and B.Frazer-Reid, *Tetrahedron Lett.,* 1988, **29**, 979.

6)   S.Ozaki, Y.Kondo, H.Nakahira, S.Yamaoka and Y.Watanabe, *Tetrahedron Lett.,* 1987, **28**, 4691.

7)   G.Baudin, B.I.Glanzer, K.S.Swaminathan and A.Vasella, *Helv.Chim.Acta.,* 1988, **71**, 1367.

8)   H.W.Lee and Y.Kishi, *J.Org.Chem.,* 1985, **50**, 4402.

9)   C.E.Dreef, R.J.Tuinman, C.J.J.Elie, G.A.van der Marel and J.H.van Boom, *Recl.Trav.Chim.Pays-Bas,* 1988, **107**, 395.

10) J.L.Meek, F.Davidson and F.W.Hobbs, *J.Am.Chem.Soc.,* 1988, **110**, 2317.

11) Y.Watanabe, T.Shinohara, T.Fujimoto and S.Ozaki, *Chem.Pharm.Bull.,* 1990, **38**, 562.

12) Y.Watanabe, A.Oka, Y.Shimizu and S.Ozaki, *Tetrahedron Lett.,* 1990, **31**, 2613.

13) Y.Watanabe, T.Fujimoto, T.Shinohara and S.Ozaki, *J.Chem.Soc.Chem.Commun.,* 1991, 428.

14) H.A.J.Carless and K.Busia, *Tetrahedron Lett.,* 1990, **31**, 3449.

15) Y.Watanabe, M.Mitani, T.Morita and S.Ozaki, *J.Chem.Soc.Chem.Commun.,* 1989, 482.

Chapter 8

Analogues of the Inositol Phosphates :
Chemical Synthesis and Biological Activity

8.1 Introduction

Due to the fundamental second messenger role of inositol 1,4,5-trisphosphate (1,4,5 -IP_3) outlined in Chapter 2, most of the synthetic studies reported to date have been concerned with the preparation of analogues of this molecule. A combination of the pivotal position of inositol 1-phosphate (I-1-P) in the metabolic sequence used for the recycling of 1,4,5-IP_3, and the tempting hypothesis that the enzyme which dephosphorylates I-1-P to free inositol is the primary site of action for the well-known pharmacological activity of lithium in the treatment of manic depression, discussed in Chapter 2, have resulted in the preparation of analogues of I-1-P as potential lithium mimics. Synthesis of analogues of inositol bisphosphates, and of inositol 1,3,4,5-tetrakisphosphate have also appeared, mainly representing the application of methods developed for 1,4,5-IP_3 analogues to available synthetic intermediates, rather than targeted syntheses.

For the purposes of this chapter, I have taken analogues to mean compounds synthesised with the aim of producing molecules which interact with receptors or enzymes which would normally recognise the inositol phosphates, thus mimicking or blocking the biological activity of the naturally occurring molecules.

8.2 Molecular Modifications

The modifications reported to date fall into two broad classes, those designed to mimic or replace the phosphate group(s), and those concerned with modification of the inositol hydroxyl groups.

Phosphate mimics/replacements have been prepared with the following goals in mind: 1) to increase stability to the metabolic enzymes (phosphatases) which degrade the natural materials, resulting in a longer biological half-life; and 2) to decrease polarity in order to facilitate the entry of the analogues into intact cells and into the central nervous system (CNS). Analogues displaying both of these properties, whilst retaining biological activity, would be highly desirable. Classic modifications of the phosphate group include conversion into the non-hydrolysable phosphonate analogue, replacing P=O by P=S to reduce the rate of enzymic hydrolysis, and replacing the acidic phosphate OH groups by other functions,*eg* esterification, *Fig 8-1*.

Replacement or modification of the inositol hydroxyl groups has been undertaken mainly in order to obtain analogues which: 1) cannot be metabolised by the normal routes seen for the

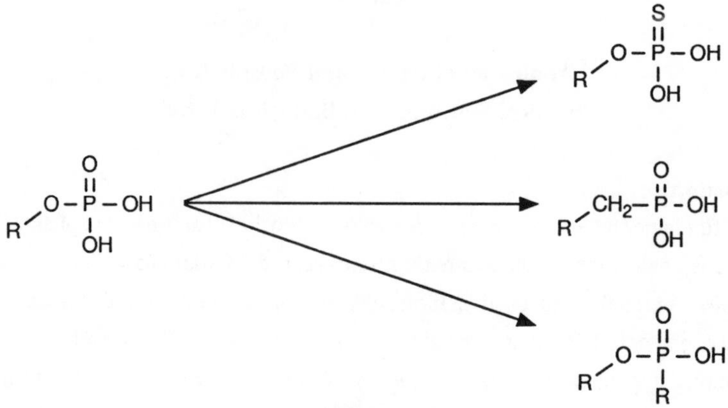

Fig 8 - 1

natural materials (*eg via* kinases), and thus have longer biological half-lives; and 2) to decrease polarity of the overall molecule for the reasons given above. Application of traditional medicinal chemistry methods leads to the replacement of OH by H (simple deletion), OR (maintaining the ability to accept a hydrogen bond), F or F_2 (similar electronic character to OH but more hydrophobic), alkyl groups (possible hydrophobic interactions, and possible antagonist activity) or inversion of the relative stereochemistry of the R-CH(OH)-R system, *Fig 8-2*.

Thus the overall goals of the two approaches are convergent, and analogues containing both phosphate mimics and hydroxyl modifications represent rational and important targets, and could provide potential drug candidates.

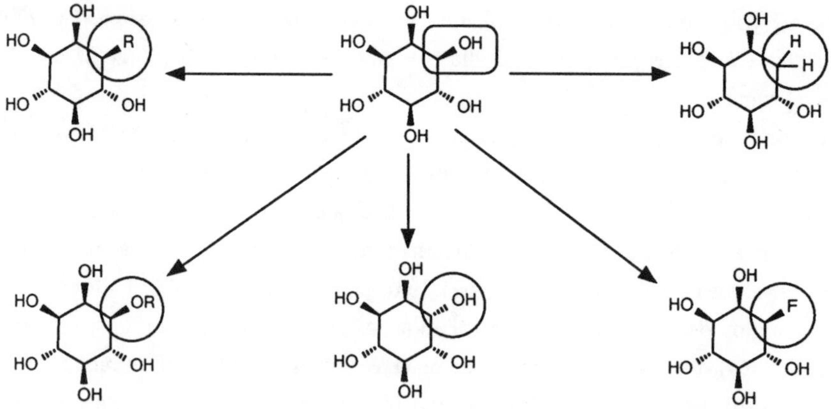

Fig 8 - 2

8.3 Analogues of Inositol 1-Phosphate

The enzyme inositol monophosphatase has been chosen as a target for inhibition due to its pivotal role in the recycling of the inositol phosphates, and the probability that inhibition of this monophosphatase represents the primary site of action of the anti-manic drug lithium, see Chapter 2 for a detailed discussion. Thus the synthesis of analogues of the natural substrates which could act as competitive inhibitors of the enzyme is an attractive approach to producing a replacement for lithium therapy in manic depression.

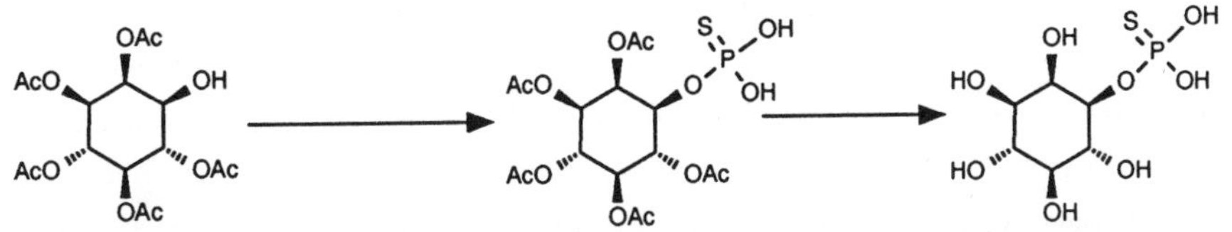

Fig 8 - 3

Racemic inositol 1-phosphorothioate (3) has been prepared from the pentaacetate (1), *Fig 8-3, via* treatment with thiophosphoryl chloride giving the partially protected intermediate (2), and subsequent deprotection with ammonia in methanol.[1] Under these conditions the free phosphorothioate precipitates directly from the methanolic solution, and can be isolated in *ca* 20% yield. The authors provided no biological data for this compound. In a more recent

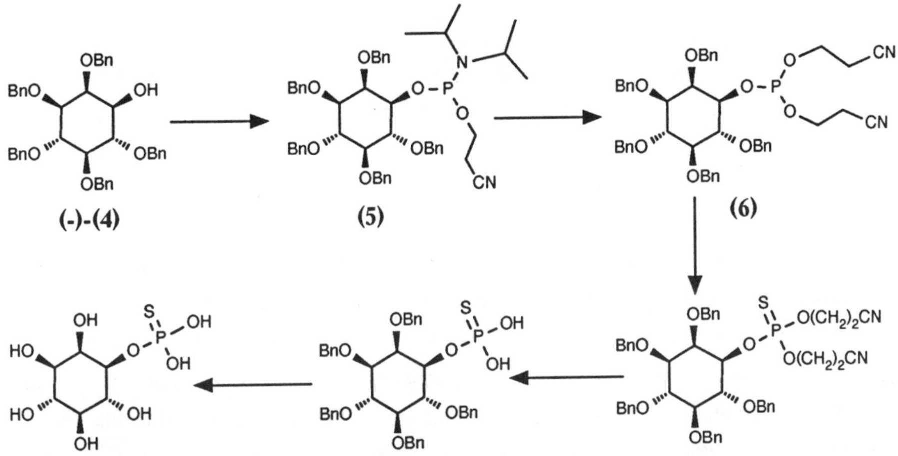

Fig 8 - 4

approach,[2,3] more modern techniques were applied to the synthesis of both the racemic material and the individual enantiomers. The individual enantiomers of pentabenzyl inositol *eg* (-)-(4) were prepared by the previously reported resolution using the camphanate esters,[4] and the synthetic route is illustrated for a single enantiomer in *Fig 8-4*. Treatment of (-)-(4) with 2-cyanoethyl N,N-diisopropylchlorophosphoramidite in CH_2Cl_2 gave the intermediate phosphoramidite ester (5), *Fig 8-4*, which was converted into the phosphite triester (6) by reaction with 3-hydroxypropionitrile *in situ*. Treatment of the crude triester with sulphur in pyridine then provided the phosphorothioate triester (7), in 75% overall yield from the alcohol (4).[5,6] Base-catalysed β-elimination of the phosphate moiety from both of the cyanoethyl groups gave the free phosphorothioate (8), and subsequent hydrogenolysis of the benzyl ethers gave the enantiomers of inositol 1-phosphorothioate, *eg* (-)-(9) from alcohol (-)-(4), in *ca* 60% overall yield. Both of the enantiomeric phosphorothioates acted as weak competitive inhibitors of the hydrolysis of inositol 1-phosphate by inositol monophosphatase, with an observed K_i of 1.0 mM for each enantiomer. Both of the enantiomeric phosphorothioates were substrates for the monophosphatase, but were hydrolysed more slowly than the parent phosphates. In a related approach the same authors synthesised

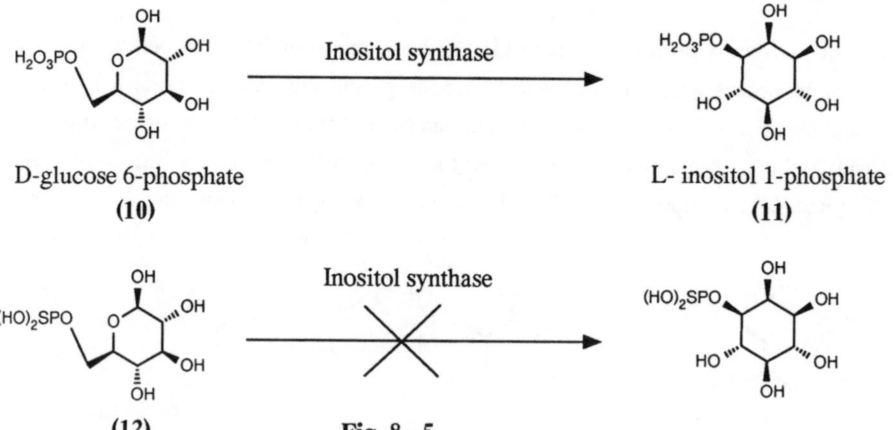

D-glucose 6-phosphate **(10)** → Inositol synthase → L- inositol 1-phosphate **(11)**

(12)   Inositol synthase (crossed out)

Fig 8 - 5

D-glucose 6-phosphorothioate (12) using similar methodology, and attempted to convert this into L-inositol 1-phosphorothioate *via* the enzyme inositol synthase, *Fig 8-5*. Although this enzyme rapidly converts D-glucose 6-phosphate (10) into L-inositol 1-phosphate (11), see Chapter 1, incubation with D-glucose 6-phosphorothioate under identical conditions produced no L-inositol 1-phosphorothioate. This curious result suggests that the phosphate group plays a previously unrecognised important and specific role in this enzymic conversion.

Initial efforts to synthesise the phosphonate isostere (23) of inositol 1-phosphate as a potential non-hydrolysable competitive inhibitor of inositol monophosphatase met with frustration,[7] as treatment of the ketone (13) with the anion of the methylene bisphosphonate shown, *Fig 8-6*, under Wittig-Horner conditions gave only the phenol (14), and attempted

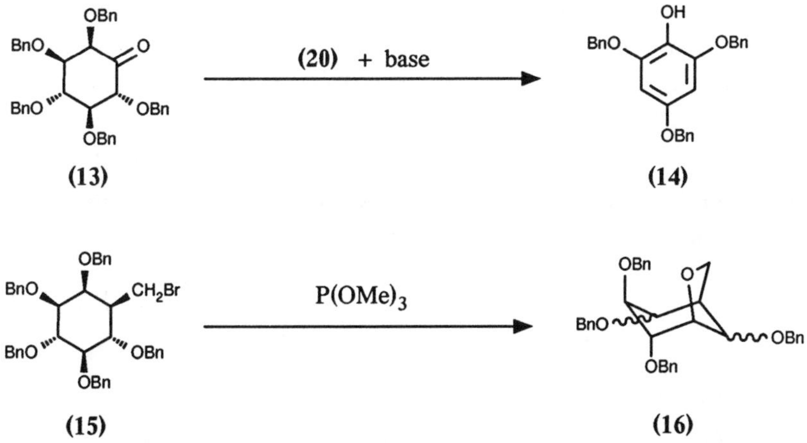

Fig 8 - 6

Arbusov reaction of the bromide (15) with trimethyl phosphite gave only the bicyclic compound (16). These results suggested that a rigid analogue of ketone (13) was desirable. Selective monobenzylation of the known 1,2;4,5-bisisopropylidine acetal[8] gave the monobenzyl ether (17) as the major product, *Fig 8-7*. Silylation of the free hydroxyl group, followed by hydrogenation gave the isomeric protected derivative (18), which could be

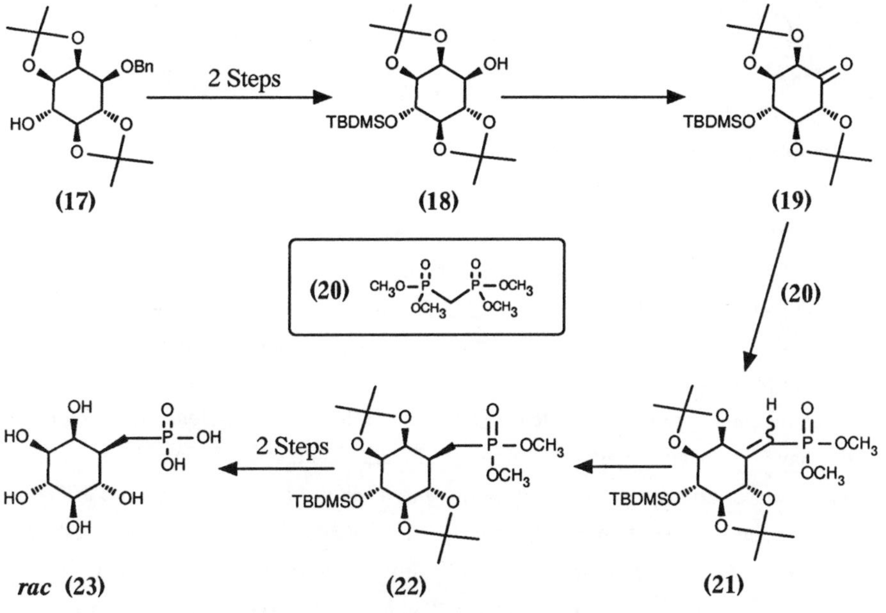

Fig 8 - 7

oxidised to the ketone (19). Treatment of (19) with the anion of the methylenephosphonate

(20) then provided the unsaturated phosphonate (21), as a *cis/trans* mixture. Hydrogenation of this material over platinum resulted in exclusive addition of hydrogen from the alpha face, giving the desired *myo*-inositol stereochemistry in the product (22). Transesterification of the methyl phosphonate with bromotrimethylsilane, followed by acidic hydrolysis of the acetal groups then gave racemic inositol 1-phosphonate, *rac*-(23). No biological data was reported with the synthesis of this compound, however it has subsequently been disclosed to be devoid of enzyme inhibitory activity.[9]

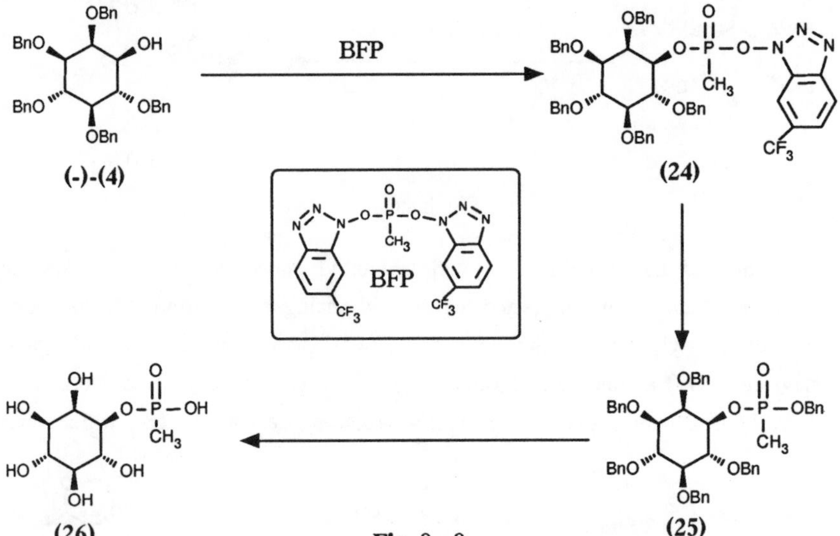

Fig 8 - 8

The methylphosphonate analogue of D-inositol 1-phosphate has been prepared, *Fig 8-8*, from the alcohol (-)-(4) *via* a bifunctional phosphonylating agent .[10] Reaction of (-)-(4) with the reagent shown at 20°C gave the intermediate ester (24). Subsequent *in situ* reaction with benzyl alcohol provided the fully protected methylphosphonate (25), which on deprotection by hydrogenation gave D-inositol 1-methylphosphonate, (26). The authors report no biological data for this compound, but speculate optimistically that it may be an inhibitor of inositol monophosphatase.

The medicinal chemistry group at Merck Sharp and Dohme's neuroscience research centre in the UK have reported significant progress in the synthesis of inhibitors of inositol monophosphatase, involving both the development of phosphate mimics and modifications of the inositol ring. Inositol monophosphatase hydrolyses both enantiomers of inositol 1-phosphate, and both enantiomers of inositol 4-phosphate, in addition to certain β-glycerophosphate derivatives, and 2'-AMP. The Merck group questioned the role of the 2- and 6- hydroxyl groups in the recognition of inositol 1-phosphate by the enzyme, and synthesised the 2- and 6- deoxyinositol 1-phosphates to address this question.[11]

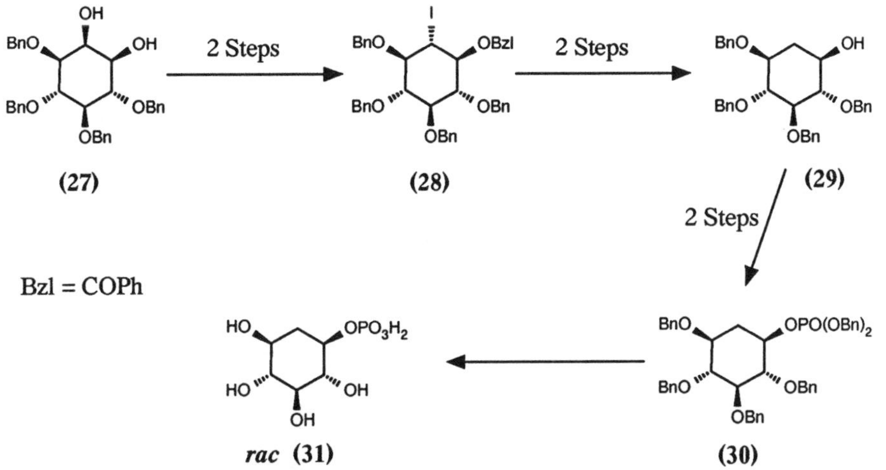

Fig 8 - 9

Benzylation of the well-known 1,2-cyclohexylidine acetal, followed by hydrolysis of the acetal group gave the diol (27), *Fig 8-9*.[12] Formation of the cyclic stannylene of (27) using dibutyltin oxide, followed by reaction with benzoyl chloride gave the 1-benzoyl derivative, which was halogenated using iodine and triphenyl phosphine in the presence of imidazole to provide the inverted iodide (28). Reductive dehalogenation with tributyltin hydride in the presence of a radical initiator, followed by hydrolysis of the benzyl ester then gave the alcohol (29). Phosphorylation of (29) with diphenylchlorophosphate, followed by transesterification with the anion of benzyl alcohol produced the fully protected deoxyphosphate (30), which on deprotection by hydrogenolysis gave racemic 2-deoxyinositol 1-phosphate, *rac*-(31).

Synthesis of the 6-deoxy compound followed a similar strategy, starting from the 2,3;4,5-biscyclohexylidine acetal (32), *Fig 8-10*. Benzoylation of (32) in pyridine gave the 1-benzyl derivative as the major product, and halogenation as in the previous synthesis gave the inverted iodide (33). Reductive dehalogenation as before, followed by hydrolysis of the benzyl ether provided the alcohol (34). Phosphorylation of (34) with diphenylchlorophosphate, followed by chromatography on silica gel resulted in cleavage of the less stable acetal group, giving the diol (35) in 65% yield. Hydrogenation of (35) over a platinum catalyst removed the phenyl phosphate esters, and resulted in concomitant acetal cleavage, giving racemic 6-deoxyinositol 1-phosphate, *rac*-(36).

Biological testing of these two deoxy compounds revealed that both *rac*-(31) and *rac*-(36) were competitive inhibitors of the hydrolysis of inositol 1-phosphate by inositol monophosphatase, with equal IC_{50}s of *ca* 70 µM. In addition incubation with the pure

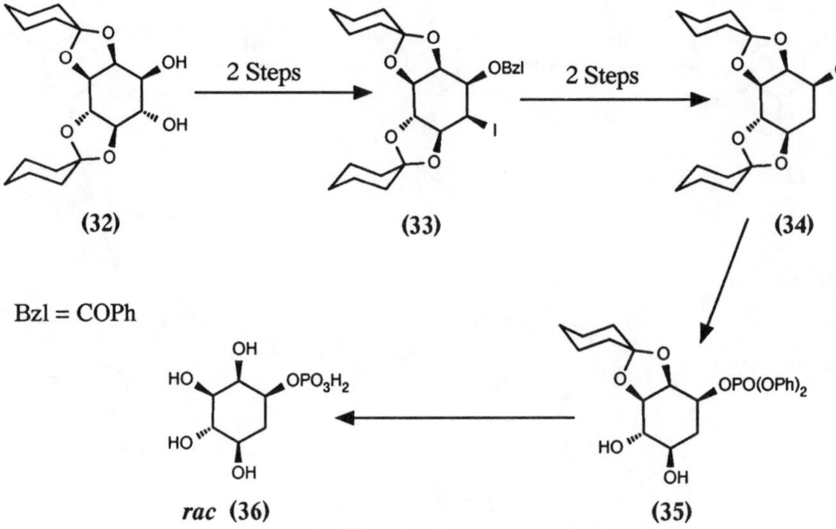
2 Steps
2 Steps
(32)
(33)
(34)
Bzl = COPh
rac (36)
(35)

Fig 8 - 10

enzyme released no inorganic phosphate, demonstrating that the compounds were not substrates for the enzyme. These results indicated that both flanking OH groups were necessary for efficient phosphate hydrolysis by the monophosphatase, and prompted the resolution of the 2-deoxy derivative to further probe the enzymic mechanism.

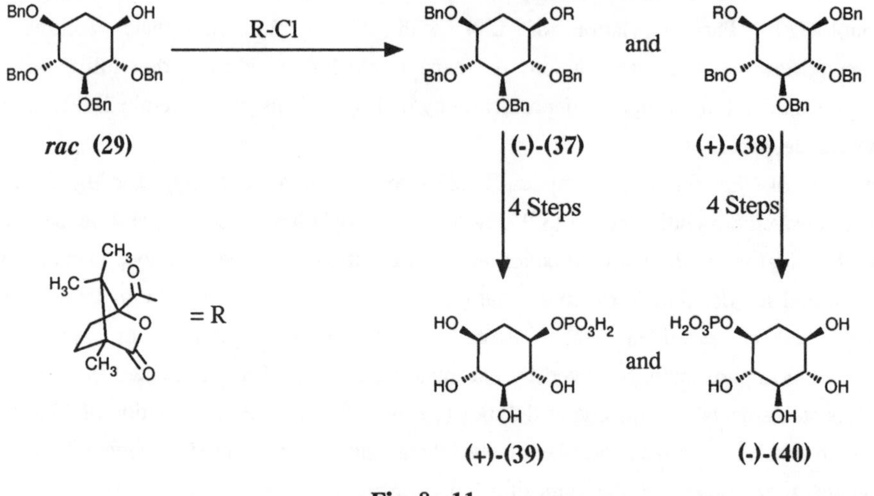
R-Cl
and
rac (29)
(-)-(37)
(+)-(38)
4 Steps
4 Steps
= R
and
(+)-(39)
(-)-(40)

Fig 8 - 11

Resolution of alcohol (29) was achieved by conversion to the diastereomeric camphanate esters (37) and (38),[4] separation of the diastereomers by chromatography, and hydrolysis of the camphanates. The absolute configuration of the camphanate (+)-(38) was established by X-ray crystallography as *S,R,S,R,R* as depicted in *Fig 8-11*. The enantiomeric alcohols derived from these camphanates were then converted into the enantiomers of 2-deoxyinositol 1-phosphate (+)-(39) and (-)-(40), having the absolute configurations shown, by the methods used in the racemic series. The biological activity of these enantiomers suggested that the two flanking hydroxyls play discrete and different roles in the recognition and hydrolysis of inositol 1-phosphate by the monophosphatase. Thus (-)-(40) was a competitive inhibitor of monophosphatase activity, with IC_{50} = 50 µ*M*, and was not a substrate for phosphate hydrolysis by the pure monophosphatase. In contrast, (+)-(39) was only a very weak inhibitor of the monophosphatase, but was slowly hydrolysed by the pure enzyme with V_{max} = 78% of that observed for inositol 1-phosphate, and K_m = 1.3 m*M*. Based on these results, and the results for the racemic deoxy compounds, a model was proposed for substrate recognition which involves one flanking hydroxyl in a binding interaction with the active site, whilst the second hydroxyl contributes little to the overall binding energy, but is involved in the mechanism of phosphate hydrolysis, and thus is required for substrate activity. This proposal is represented in *Fig 8-12*, where it can be seen that (+)-(39) possesses the mechanistic hydroxyl, but not the binding hydroxyl, and thus is a weak substrate for the enzyme. In contrast (-)-(40) is a good competitive inhibitor of the enzyme, but is not hydrolysed, reflecting the presence of the binding hydroxyl coupled with the absence of the mechanistic hydroxyl.

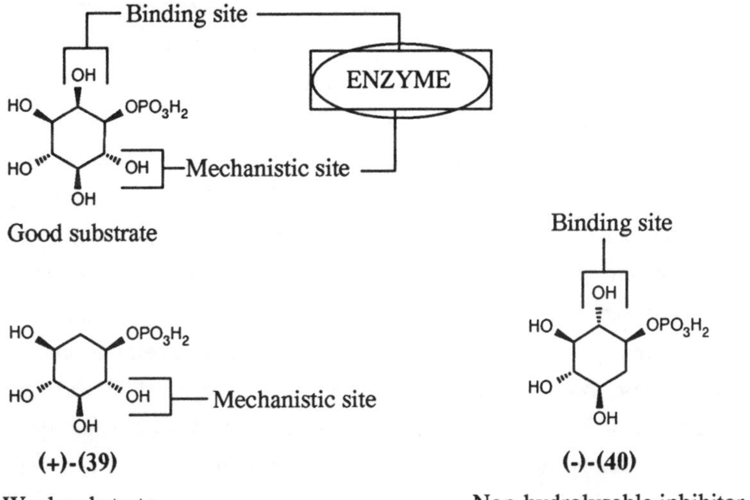

Fig 8 - 12

Molecular modelling studies of the relationship between the enantiomers of inositol 1-phosphate suggested that perhaps the 3- and 5- hydroxyl groups were not directly involved in interactions with the enzyme.[13] This led to the synthesis of the 3,5-dideoxy derivatives of (36) and (31), *ie* (46) and (51), *Fig 8-13* and *Fig 8-14*. Epoxidation of cyclohexenol (41), followed by benzylation of the free hydroxyl group gave the *cis*-epoxide (42), *Fig 8-13*.

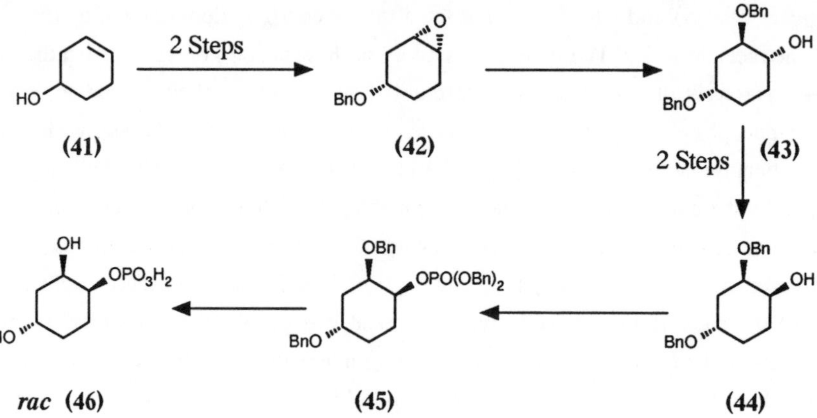

Fig 8 - 13

Treatment of (42) with benzyl alcohol in refluxing toluene containing alumina[14] resulted in regioselective epoxide opening, giving alcohol (43). Swern oxidation of (43) to the corresponding ketone, followed by reduction using L-Selectride provided the inverted alcohol (44). Phosphorylation of (44) using tetrabenzylpyrophosphate then gave the fully protected derivative (45), which on hydrogenolysis gave the target compound *rac*-(46). A similar strategy, this time involving inversion of the hydroxyl at position 4, was followed for the isomeric compound (51), *Fig 8-14*. Epoxidation of cyclohexenol (41), followed by protection as the *p*-methoxybenzyl ether gave (47). Regioselective epoxide opening using benzyl alcohol as before, followed by protection of the free hydroxyl group by silylation gave (48). Removal of the *p*-methoxybenzyl group and oxidation of the resulting alcohol to the ketone then gave (49). Selective reduction using sodium borohydride, followed by benzylation of the free hydroxyl group, and cleavage of the silyl group provided the alcohol (50). Finally phosphorylation and deprotection of (50) as before gave the target compound, *rac*-(51).

Both of these cyclohexane triol phosphates were competitive inhibitors of inositol monophosphatase, *rac*-(51) having $IC_{50} = 90$ μM whilst *rac*-(46) was some 10-fold more potent with $IC_{50} = 7$ μM. The individual enantiomers of (46) were then synthesised from the enantiomers of alcohol (44), obtained *via* the diastereomeric camphanate esters of (44). The absolute stereochemistry of one of the camphanate esters was determined by X-ray crystallography as before, and this allowed the absolute configuration of the final phosphates (-)-(52) and (+)-(53) to be assigned as 1*S*,2*R*,4*S* and 1*R*,2*S*,4*R* respectively as shown in *Fig 8-15*. In agreement with the results obtained for the 2-deoxy compounds, see *Fig 8-12*,

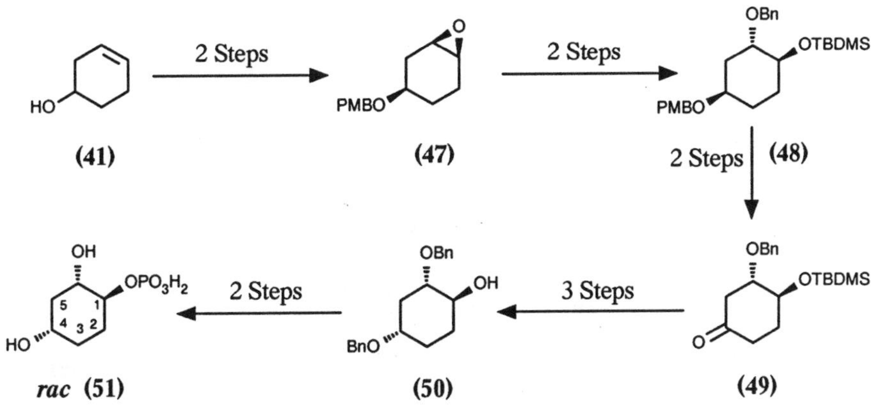

Fig 8 - 14

(-)-(52) proved to be a potent competitive inhibitor of the monophosphatase IC$_{50}$ = 3 μM, whilst (+)-(53) was a weak substrate with little inhibitory potency. From these results the authors concluded that the 3- and 5- hydroxyl groups in both the inositol monophosphates and their deoxy derivatives are not required for recognition by the enzyme, and that (-)-(52) represents the minimum essential structural features necessary for enzyme recognition, in the optimum stereochemical configuration.

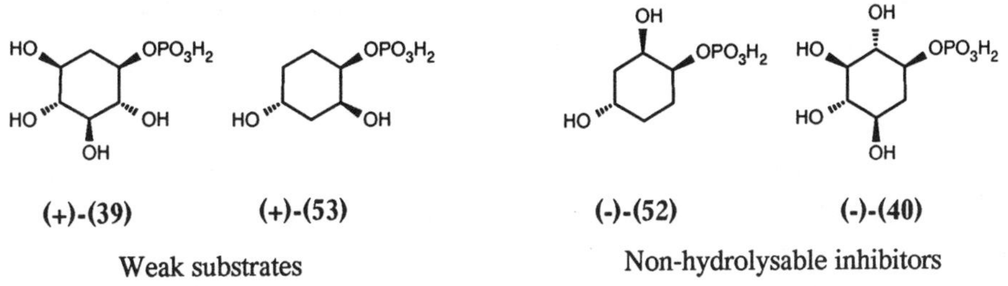

Fig 8 - 15

Further molecular modelling comparisons between the structure of (46) and other substrates, in particular 2'-AMP, suggested that space to accommodate bulky groups was present in the enzyme active site, corresponding to substitution at the vacant 6-position of (46).[15] Treatment of epoxide (42) with acetic acid in the presence of an alumina catalyst gave the acetate (54), *Fig 8-16*. Formation of the methanesulphonate ester of (54), followed by treatment with base resulted in hydrolysis of the acetate ester, and ring closure to the isomeric epoxide (55). Treatment of (55) with sodium phenylselenide led to the intermediate selenide (56) by regiospecific epoxide opening, and in-situ oxidation to the selenoxide and subsequent elimination gave the allylic alcohol (57). Directed epoxidation of (57), followed

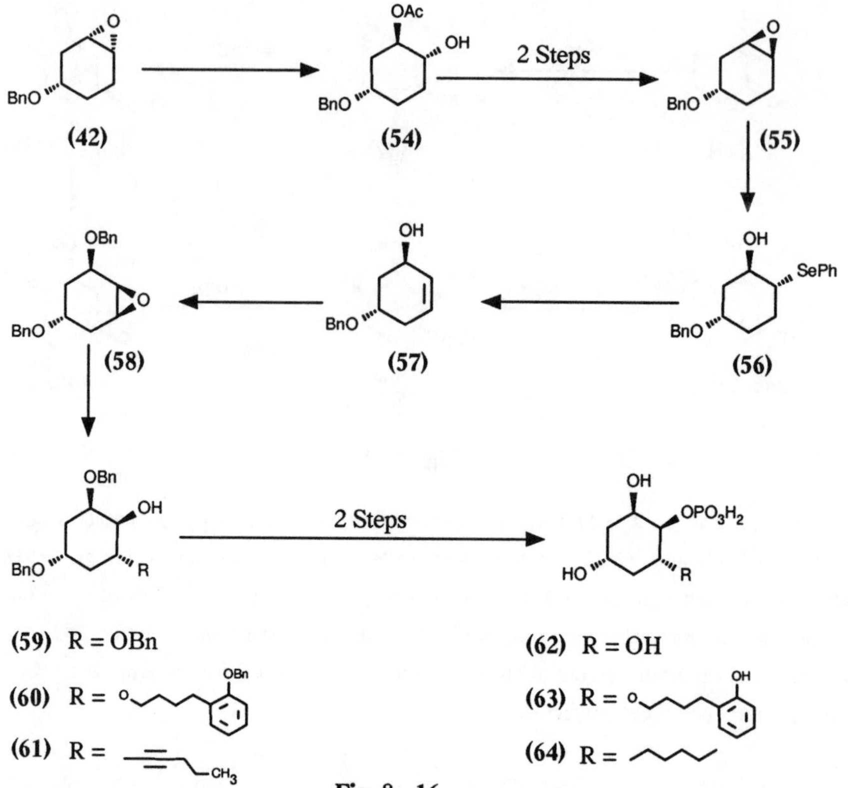

Fig 8 - 16

by benzylation of the free hydroxyl group provided the key intermediate (58). Regiospecific epoxide opening using both oxygen and carbon nucleophiles proved possible, leading to the 6-substituted alcohols (59)-(61). Phosphorylation and deprotection of these derivatives then gave the racemic 6-substituted analogues (62)-(64).

The hydroxy substituted compound (62) proved to be a very good substrate for the monophosphatase, having an affinity greater than any of the natural substrates. This indicates that the detrimental effects of the 3- and 5- hydroxyl groups previously observed in the inhibitors described above also occur in substrates. The *n*-pentyl substituted compound (64) was a competitive inhibitor of monophosphatase activity with $IC_{50} = 3$ μ*M*, and as expected was not a substrate for the enzyme. Thus the bulk of the *n*-pentyl substituent was well tolerated in the enzyme active site. Introduction of the phenol-containing butyl chain in (63) however gave an inhibitor with $IC_{50} = 70$ n*M* , or *ca* 100-fold higher potency than the unsubstituted analogue (46). Clearly in this case the side chain is providing significant extra binding affinity. The 6-substituted compound (63) represents the most potent inhibitor of inositol monophosphatase disclosed to date.

The above strategy has led to potent, specific inhibitors of inositol monophosphatase, but the *in-vivo* activity of these compounds is compromised by their hydrolysis by ubiquitous non-specific phosphatases, and their lack of penetration into intact cells. This led the Merck group to seek replacements for the phosphate functionality in these molecules.[9] Replacement of the phosphate group in (46) by an isosteric phosphonate gave (65), which in parallel with the phosphonate derived from inositol 1-phosphate, (23), showed no inhibition of monophosphatase activity, *Fig 8-17*. Screening studies led to the identification of hydroxymethylenebisphosphonic acid (66) as a competitive inhibitor of monophosphatase activity $K_i = 0.18$ m*M*, whereas the analogue (67) was inactive, *Fig 8-17*. Combining the

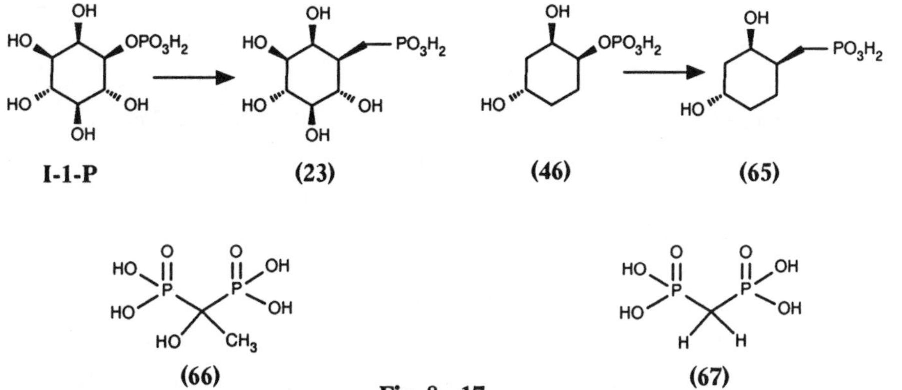

I-1-P   (23)   (46)   (65)

(66)   (67)

Fig 8 - 17

structural features of (66) and (46) led to the synthesis of the bisphosphonate (71), *Fig 8-18*.[9] Reaction of the anion derived from (44) with the trifluoromethanesulphonate ester of diethyl

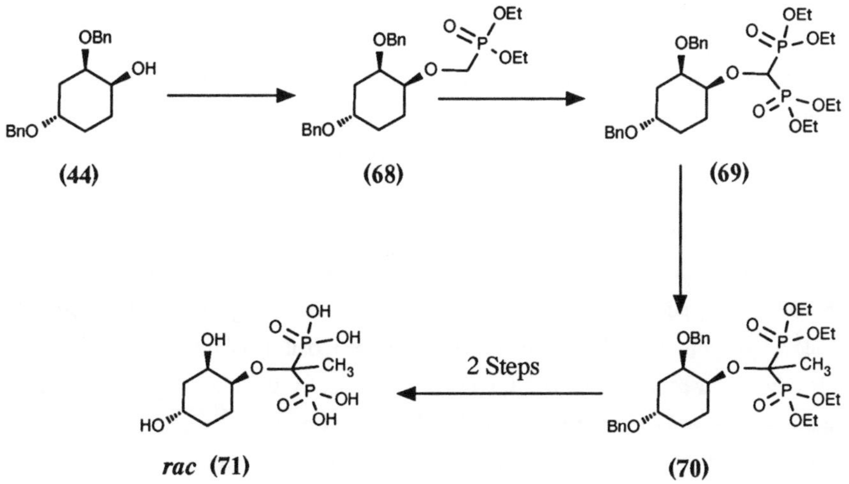

(44)   (68)   (69)

rac (71)   2 Steps   (70)

Fig 8 - 18

hydroxymethylphosphonate [16] gave the phosphonate (68). Treatment of (68) with base, followed by diethyl chlorophosphate then gave the bisphosphonate (69), which could be deprotonated and alkylated with methyl iodide to provide the fully protected compound (70). Sequential deprotection using bromotrimethylsilane followed by hydrogenolysis gave the target compound, rac-(71). The bisphosphonate (71) was a potent competitive inhibitor of monophosphatase activity, $K_i = 2.5\ \mu M$, with an affinity similar to that of the corresponding phosphate. A number of modifications of the bisphosphonate group in (71), including

(72)	(73)	(74)
$K_i = 7.4\ \mu M$	$K_i = 1.2\ \text{m}M$	$K_i = 0.7\ \text{m}M$

Fig 8 - 19

removal of the methyl group (72), removal of one of the phosphonate groups (73), and replacement of the ether oxygen by a methylene group (74) all led to reductions in potency. The considerable loss of potency observed for (74), coupled with the lack of activity seen for the simple phosphonates (23) and (65) suggests that the oxygen at the inositol C-1 position plays a substantial role in enhancing binding to the active site of the monophosphatase. This interaction may reflect an interaction of the same oxygen atom in substrates with a Lewis acid group in the active site during phosphate hydrolysis. Resolution and X-ray crystallography as before established that the absolute configuration of the more active enantiomer of (71) corresponded to the absolute configuration of the parent phosphate inhibitor (-)-(52).

Despite the significant progress reported in the synthesis of inhibitors of inositol 1-phosphatase, the fundamental problem of the lack of penetration of these compounds into intact cells, and eventually into the CNS *in vivo*, remains unsolved.

8.4 Analogues of Inositol Bisphosphates

The first reported analogues of the naturally occurring inositol bisphosphates were the phosphorothioate analogues of inositol 1,4- and 4,5- bisphosphates.[17,18] Both of these analogues were synthesised from the known protected inositols (75) and (76) by applying the methods developed by Sinha et al[19] for DNA synthesis by the phosphite method, *Fig 8-20*. Treatment of the diols with 2-cyanoethyl N,N-diisopropylchlorophosphoramidite gave the phosphite diesters, which on reaction with 3-hydroxypropionitrile *in situ* gave the phosphitetriesters,(77) and (78), *cf Fig 8-4*. Treatment of the crude triesters (77) and (78)

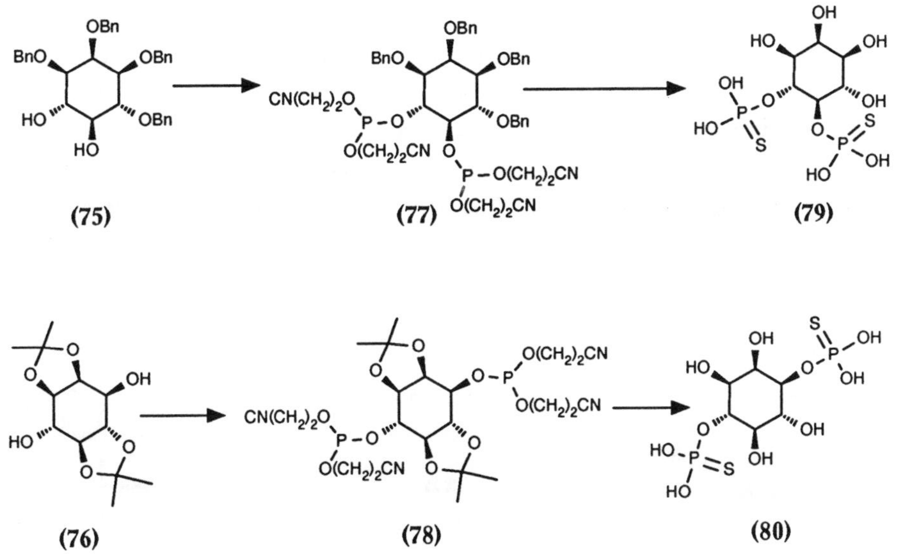

Fig 8 - 20

with sulphur in pyridine then provided the protected phosphorothioate triesters, which on deprotection gave the phosphorothioates *rac*-(79) and *rac*-(80). The authors reported that these molecules were more stable than the corresponding bisphosphates to hydrolysis by phosphatases, but no other biological data.[17]

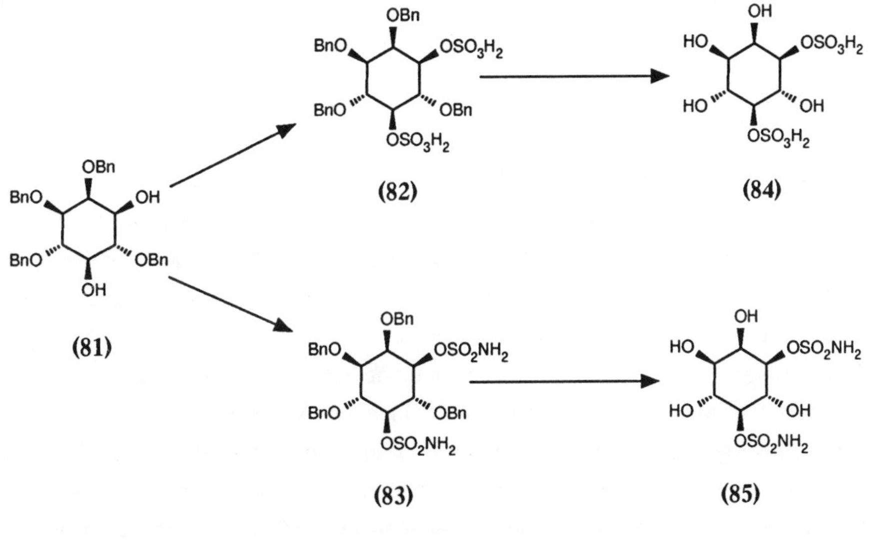

Fig 8 - 21

In an attempt to obtain less polar derivatives, the bissulphated and bissulphamoylated analogues of the non-naturally occurring inositol 1,5-bisphosphate have been prepared.[20] The key racemic diol (81), used in the synthesis of the bisphosphate, was prepared as outlined in *Fig 5-6*. Treatment of (81), *Fig 8-21*, with triethylamine-sulphur trioxide complex or sulphamoyl chloride and NaH in DMF gave the protected bissulphate (82) and bissulphonamide (83) respectively. Deprotection by hydrogenolysis gave the bissulphate, *rac*-(84), and bissulphonamide, *rac*-(85). The authors reported no biological data for these, at first sight rather contrived, analogues of a non-naturally occurring inositol bisphosphate.

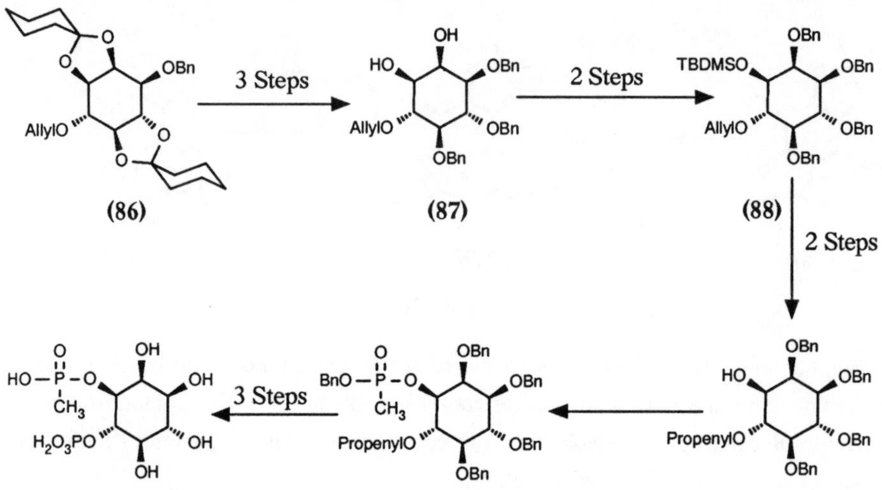

Fig 8 - 22

Replacement of one of the phosphate groups in inositol 3,4-bisphosphate by a methyl phosphonate has been reported,[21] starting from the known selectively protected bisketal (48), and using the bifunctional phosphonylating agent BFP[10] seen above in *Fig 8-8*. The fully protected inositol (86) was prepared by selective benzylation, *cf Fig 4-15*, of the more reactive 1- hydroxyl group in the 1,2;4,5-bisketal, followed by allylation of the remaining free hydroxyl group, *Fig 8-22*. Hydrolysis of the less stable ketal group, followed by benzylation of the resulting diol, and subsequent hydrolysis of the remaining ketal gave the diol (87). Regioselective silylation of (87) using the bulky *t*-butyldimethylsilyl group [22] gave exclusively the equatorial silyl derivative, which on benzylation of the remaining free hydroxyl group gave the fully protected inositol (88). Isomerisation of the allyl group into the *trans*-prop-1-enyl group, and removal of the silyl group then afforded the alcohol (89). Treatment of (89) with BFP, followed by benzyl alcohol gave the protected methyl phosphonate (90), *cf Fig 8-8*. Finally hydrolysis of the *trans*-prop-1-enyl group, phosphorylation of the resulting alcohol, and removal of all of the benzyl protecting groups by hydrogenolysis gave the 3-methylphosphonate-4-phosphate derivative, *rac*-(91). The authors reported no biological data for this derivative.

8.5 Analogues of Inositol Trisphosphates

Due to the central biological importance of the trisphosphates, and especially 1,4,5-IP$_3$, a large body of work has been published concerning modifications of these molecules. Analogues designed to have increased biochemical stability, and to inhibit the metabolism of the natural materials have been prepared. In addition analogues have been synthesised with more specific biochemical roles in mind, such as photoaffinity labelled analogues for receptor localisation and purification, caged molecules for kinetic studies, and materials designed to be used for affinity chromatography. Any classification of the reported analogues will fall short of ideal, and here I have chosen to group the syntheses by the type of chemical modification(s) involved, rather than by the biological role envisaged for the molecules. The rationale for synthesis and an outline of the biological data available will be given for each analogue, and for detailed biological activity the interested reader is encouraged to consult the original literature cited.

8.5.1 Phosphorothioates

Phosphorothioate analogues of nucleotides have proven very valuable in both mechanistic enzymology and molecular biology,[23] due to their similar biological activity, but enhanced stability towards phosphatases compared with the parent phosphates. Thus it is not surprising that the first reported analogue of an inositol phosphate was the 1,4,5-trisphosphorothioate (1,4,5-IPS$_3$).[5] In a modification of their synthesis of 1,4,5-IP$_3$ [24] Cooke *et al* treated the

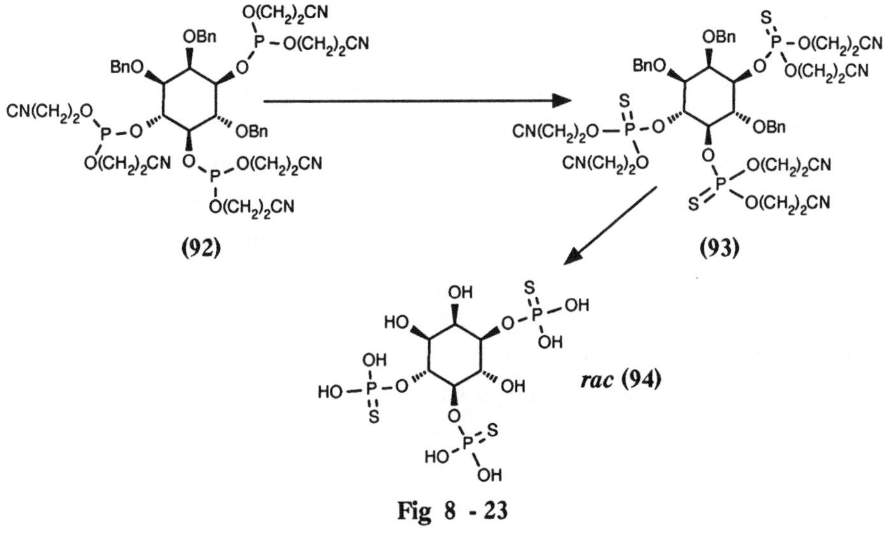

Fig 8 - 23

intermediate phosphite triester (92), *Fig 8-23*, prepared as shown in *Fig 6-2*, with sulphur in pyridine to obtain the protected trisphosphorothioate (93), which could be deprotected using sodium in liquid ammonia to racemic 1,4,5-IPS$_3$, *rac*-(94). The conversion of the racemic triol (95), prepared as shown in *Fig 6-7*, into *rac*-(94) *via* treatment with N,N-diisopropyl dibenzyl phosphoramidite, followed by phenacetyl disulphide to give (96) and deprotection using sodium in liquid ammonia has been reported, *Fig 8-24*.[25]

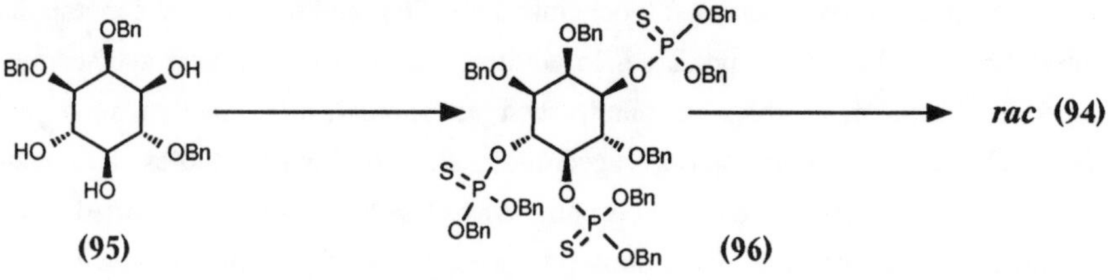

Fig 8 - 24

1,4,5-IPS$_3$ is active in binding to the specific receptor sites identified in brain[26] and hepatocytes,[27] and is a full agonist only three times less potent than the parent trisphosphate, releasing calcium from intracellular stores in *Xenopus* oocytes,[28] permeabilised Swiss 3T3 cells,[28,29] GH$_3$ cells,[29] hepatocytes,[27] pancreatic acinar cells[30] and mouse lacrimal cells.[31] As expected 1,4,5-IPS$_3$ is resistant to the 5-phosphatase that degrades 1,4,5-IP$_3$,[26,27] and thus gives rise to long-lived calcium transients in stimulated cells.[27] 1,4,5-IPS$_3$ is a potent competitive inhibitor of the 5-phosphatase-catalysed metabolism of 1,4,5-IP$_3$,[32,33,34] but unexpectedly does not bind to the 3-kinase that converts 1,4,5-IP$_3$ into 1,3,4,5-IP$_4$ and thus does not compete with 1,4,5-IP$_3$ for this enzyme.[27,33,34] A contradictory report that 1,4,5-IPS$_3$ is in fact a substrate for the 3-kinase has appeared, but no details of this study have appeared to date.[25]

In order to obtain analogues with biological activities closer to the parent 1,4,5-IP$_3$ molecule, analogues have been prepared in which only one of the phosphate groups is replaced by a phosphorothioate group. The first example of this type of molecule was the 1,4-bisphosphate-5-phosphorothioate (1,4,5S-IP$_3$) whose synthesis is shown in *Fig 8-25*.[35] Phosphorylation of the alcohol (97)[36] with bis(2,2,2-trichloroethyl) phosphorochloridate, followed by removal of the isporopylidine group gave the diol (98). Phosphorylation of this material gave a mixture of bisphosphorylated compounds, from which the crystalline isomer (99) could be obtained. Phosphitylation of (99) followed by treatment with sulphur in pyridine then gave the fully protected compound (100), which on deprotection with sodium in liquid ammonia gave racemic 1,4,5S-IP$_3$, *rac*-(101). A rather more selective approach to the synthesis of *rac*-(101),[25] based on the reaction of the protected bisphosphate (102) with N,N-diisopropyl dibenzyl phosphoramidite, followed by phenacetyl disulphide to give (103)

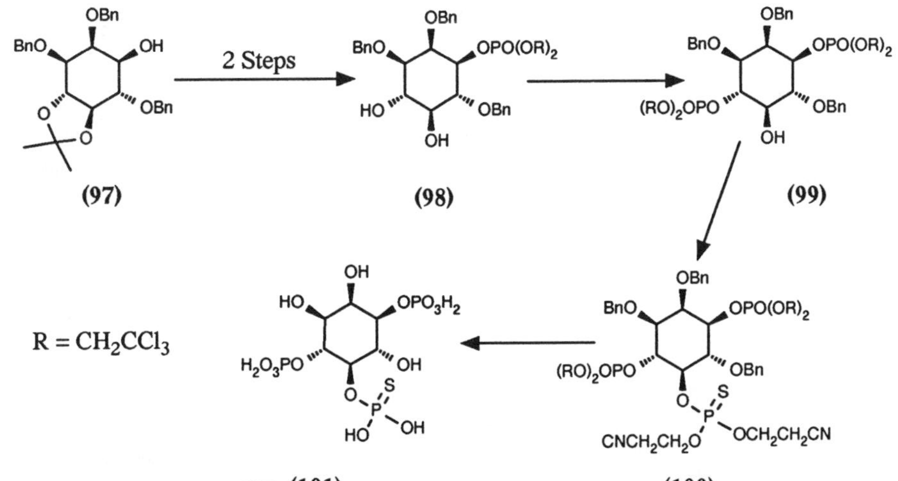

Fig 8 - 25

R = CH₂CCl₃ — R = CH_2CCl_3

and subsequent deprotection using sodium in liquid ammonia has been reported, *Fig 8-26*.
1,4,5S-IP₃ behaves in a similar fashion to 1,4,5-IPS₃, in that it binds to cerebellar receptors, stimulates the release of calcium from intracellular stores, and is an inhibitor of the 5-phosphatase.[37] In contrast to 1,4,5-IPS₃ however, 1,4,5S-IP₃ binds to the 3-kinase, and competitively inhibits the metabolism of 1,4,5-IP₃ by this route.[33] The observed slow loss of

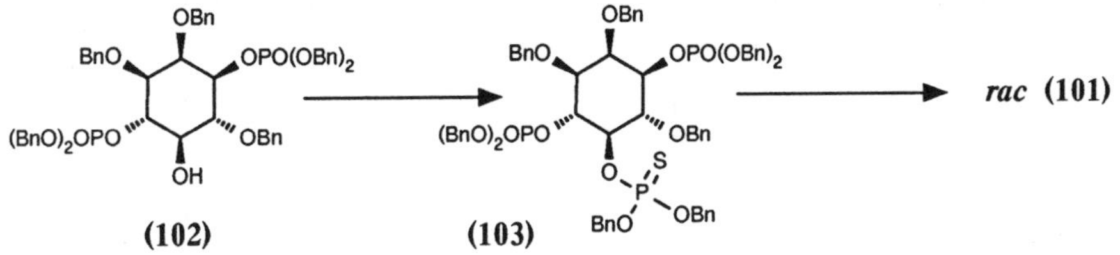

Fig 8 - 26

activity of 1,4,5S-IP₃ on incubation with the 3-kinase has been taken to indicate that the molecule is a substrate for this enzyme, presumably being phosphorylated to the 1,3,4,5-IP₄ analogue 1,3,4,5S-IP₄. Thus the different and complementary activities of these two analogues 1,4,5-IPS₃ and 1,4,5S-IP₃ may provide a useful tool to dissect the roles of 1,4,5-IP₃ and 1,3,4,5-IP₄ in calcium homeostasis.

The synthesis of the isomeric inositol 1-phosphorothioate-4,5,-bisphosphate (1S,4,5-IP₃) has been reported, and this molecule has been used for the attachment of a fluorescent "reporter" group.[38] Isomerisation of the allyl group in (104), *Fig 8-27*, to the prop-1-enyl derivative (105), followed by phosphitylation , oxidation to the phosphate level, and HgO-HgCl₂

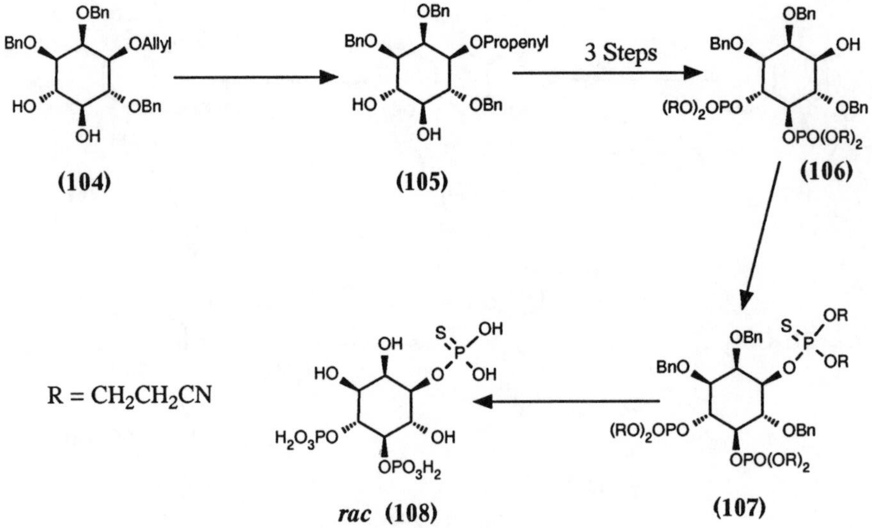

Fig 8 - 27

treatment to remove the prop-1-enyl group gave the alcohol (106). Phosphitylation of (106) followed by oxidation with sulphur in pyridine then gave the fully protected monophosphorothioate (107), which on deprotection with sodium in liquid ammonia gave racemic 1S,4,5-IP$_3$, *rac*-(108). Again the derivative has also been prepared by the phosphitylation of the bisprop-1-enyl derivative (109), obtained by isomerisation of the

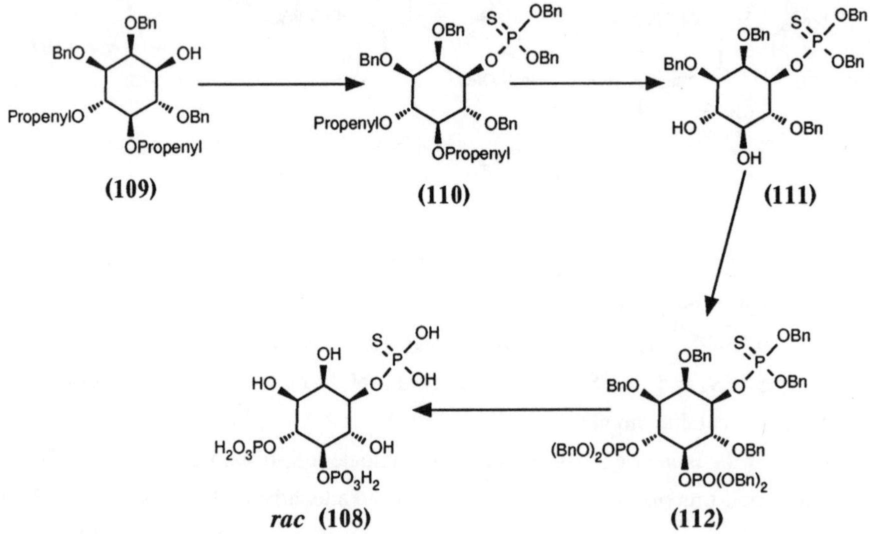

Fig 8 - 28

bisallyl derivative, with N,N-diisopropyl dibenzyl phosphoramidite, *Fig 8-28*, followed by oxidation with phenacetyl disulphide to give (110).[25] Acidolysis of the two prop-1-enyl groups then gave the diol (111), which was phosphitylated as before, and oxidised this time with *tert*-butyl hydroperoxide to give the fully protected compound (112). Deprotection of (112) with sodium in liquid ammonia then gave racemic 1*S*,4,5-IP$_3$, *rac*-(108).

Racemic 1*S*,4,5-IP$_3$ has been reported to release calcium from permeabilised cells,[38] and to be a substrate for the 3-kinase.[25] In addition the phosphorothioate has been used as a point of attachment for a fluorescent group, which may be of value in examining the interactions of 1,4,5-IP$_3$, and its analogues with proteins.

8.5.2 Phosphonates

The first phosphonate derivative of 1,4,5-IP$_3$ to be synthesised was the racemic tris-1-H-phosphonate (1,4,5-IPH$_3$).[39] Treatment of the racemic triol (95), prepared as in *Fig 6-1*, with the ammonium salt of benzyl-1-H-phosphonic acid in the presence of pivaloyl chloride gave the protected intermediate (113), *Fig 8-29*, which was debenzylated under anionic conditions (refluxing acetone and NaI) to give racemic 1,4,5-IPH$_3$, *rac*-(114). The authors report no biological data for this compound.

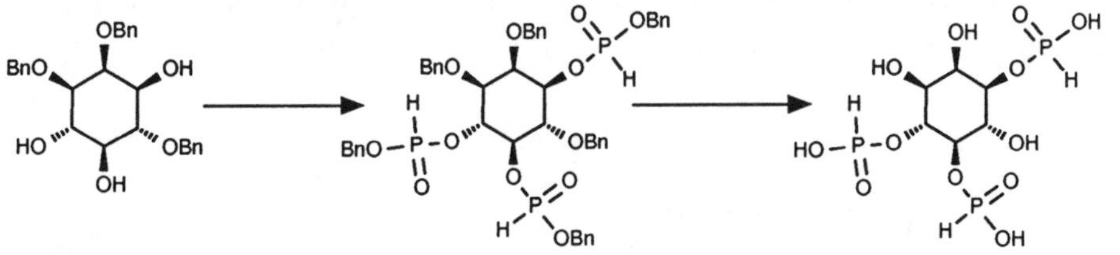

(95)   (113)   *rac* (114)

Fig 8 - 29

The 5-methylenephosphonate analogue of D-1,4,5-IP$_3$ has been synthesised from the chiral starting material (-)-quinic acid.[40] (-)-Quinic acid was converted into the intermediate olefin (115), by the same route used for the synthesis of 1,4,5-IP$_3$, outlined in *Fig 6-14*. Desilylation of (115), followed by addition of phenylselenyl bromide provided mainly the anti-Markovnikov adduct, *Fig 8-30*, which underwent smooth oxidative elimination to give the labile allylic bromide (116). Michaelis-Becker phosphorylation of (116) using dibenzylphosphite, and subsequent hydroboration of the double bond gave the diol (117). Phosphorylation of the diol (117) was accomplished by a two-step phosphitylation/oxidation protocol, and removal of the protecting groups then gave D-inositol 1,4-bisphosphate-5-methylenephosphonate, D-(118). Initial biological evaluation of D-(118) indicated that it is a long-lived D-1,4,5-IP$_3$ mimic, able to release calcium from intracellular stores.[40] A pair of related inositol 1,4-bisphosphate-5-phosphonate derivatives have been

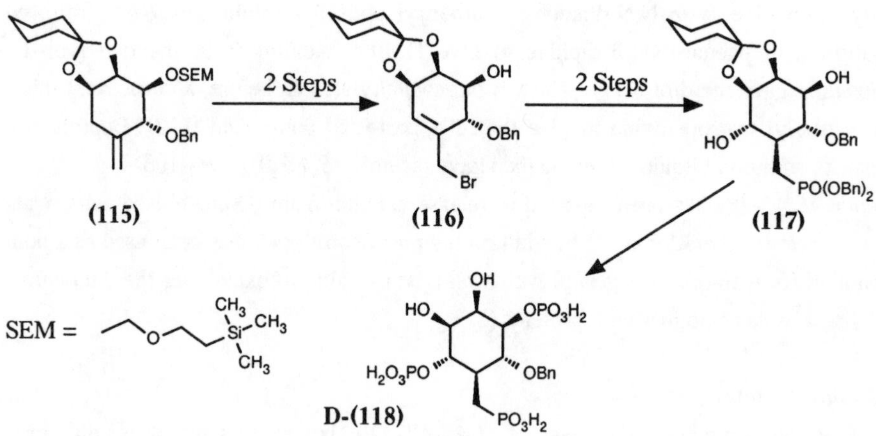

(115) (116) (117)

D-(118)

Fig 8 - 30

synthesised from the common intermediate alcohol (123), obtained as shown in *Fig 8-31*.[41] Monobenzylation of the known diol (119) followed by protection of the remaining free hydroxyl group as the *p*-methoxybenzyl ether gave (120), *Fig 8-31*. Hydrolysis of the ketal protecting group and subsequent benzylation then gave the fully protected inositol (121), which on isomerisation of the allyl groups and treatment with acid gave the diol (122).

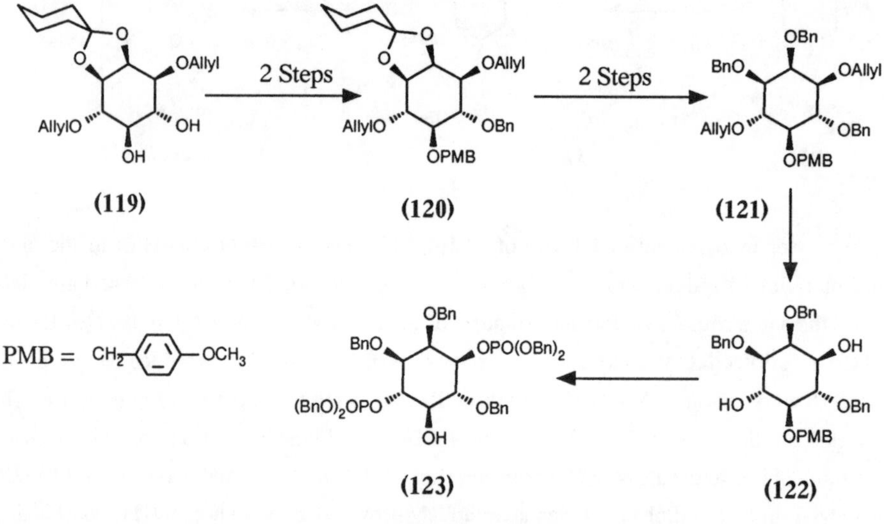

(119) (120) (121)

(123) (122)

Fig 8 - 31

Phosphitylation of (122), followed by oxidation to the bisphosphate and removal of the *p*-methoxybenzyl group then provided the key alcohol (123). Treatment of (123) with the bifunctional reagent BFP, *cf Fig 8-8*, followed by benzyl alcohol gave the fully protected methyl phosphonate (124), *Fig 8-32*, which on deprotection gave racemic inositol 1,4,-bisphosphate-5-methylphosphonate, *rac*-(125). In a similar fashion, treatment of (123) with the difluoromethyl reagent (128) gave the difluoromethylphosphonate (126), which on deprotection gave racemic inositol 1,4-bisphosphate-5-difluoromethylphosphonate, *rac*-(127). Preliminary biological results indicate that the methylphosphonate *rac*-(125) acts

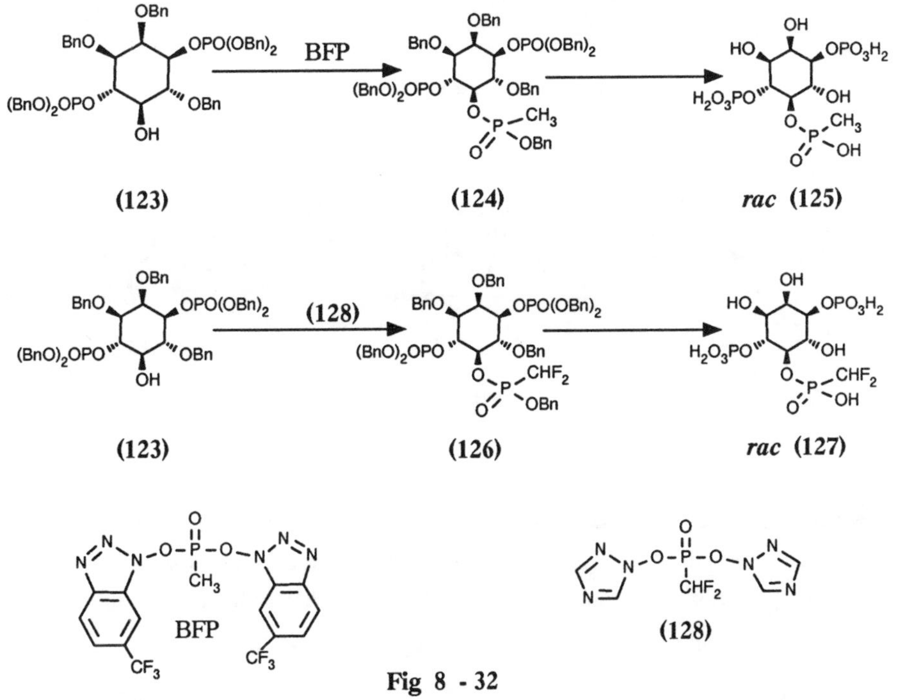

Fig 8 - 32

as an *antagonist* of 1,4,5-IP3-stimulated calcium mobilisation in permeabilised human platelets, while *rac*-(127) had no activity. If these results can be confirmed, *rac*-(125) represents the first reported 1,4,5-IP3 analogue which is an antagonist of 1,4,5-IP3-induced calcium release, and as such represents a crucial breakthrough in this area.

The same group has reported the use of this approach for the synthesis of the 3-methylphosphonate analogue of 1,3,4-IP3.[21] Using an almost identical route as shown for the conversion of (86) into (91), *Fig 8-22*, the bisallyl compound (129) was converted *via* the diol (130) into the fully protected compound (131), *Fig 8-33*. Isomerisation of the allyl groups and removal of the silyl group then provided the alcohol (132), which was converted into the protected methylphosphonate (133) using BFP as before. Hydrolysis of the propenyl groups, phosphorylation of the resulting diol, and finally deprotection by hydrogenolysis gave racemic inositol 1,4-bisphosphate-3-methylphosphonate, *rac-(134)*. The authors report no biological data for this compound.

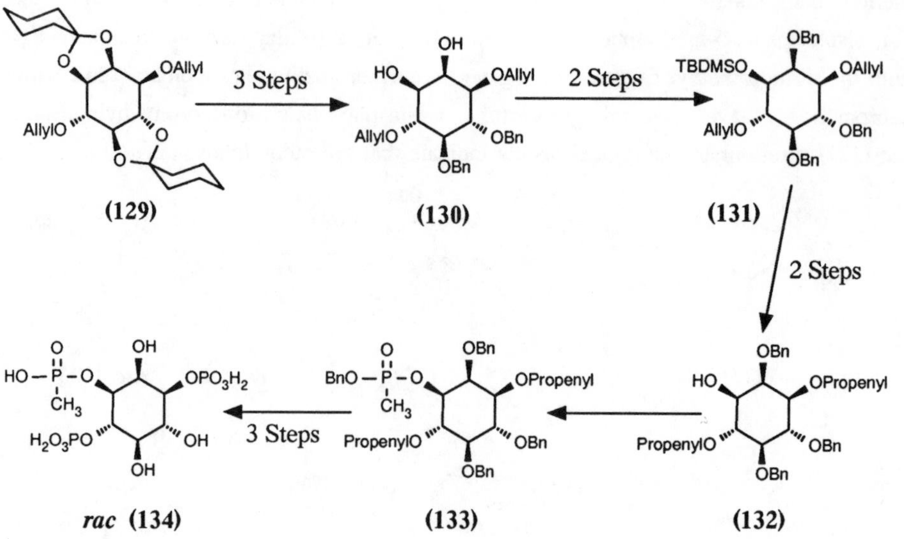

Fig 8 - 33

8.5.3 Phosphate Esters

The first report that an analogue of 1,4,5-IP$_3$ substituted at the 1-phosphate position could induce similar effects to 1,4,5-IP$_3$ concerned the biological activity of the deacylation product of phosphatidylinositol 4,5-bisphosphate, (135) *Fig 8-34*.[42] Subsequently the biological

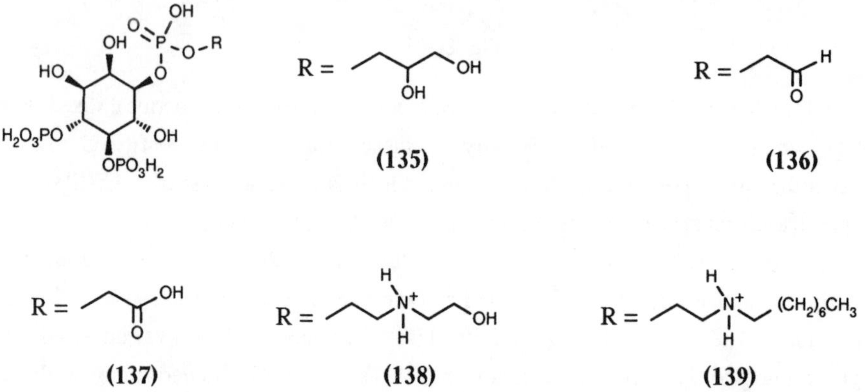

Fig 8 - 34

activity of a series of phosphate esters, (136)-(139) *Fig 8-34*, all derived from the naturally occurring phospholipid, was reported.[43] All of these compounds were full agonists at the 1,4,5-IP$_3$ receptor, and with the exception of (138) all were relatively potent in calcium release studies. Thus it was clear that substantial modification of the phosphate group at the 1- position of 1,4,5-IP$_3$ was tolerated, without loss of binding or agonist activity. It was concluded from these and other results that the 4,5-vicinal phosphate moiety is essential for calcium mobilising activity, and the 1-phosphate group, whilst not essential, provides extra binding interactions with the 1,4,5-IP$_3$ receptor. Following similar methodology the same authors have prepared photoaffinity labelled 1,4,5-IP$_3$ analogues (140) and (141), *Fig 8-35*,

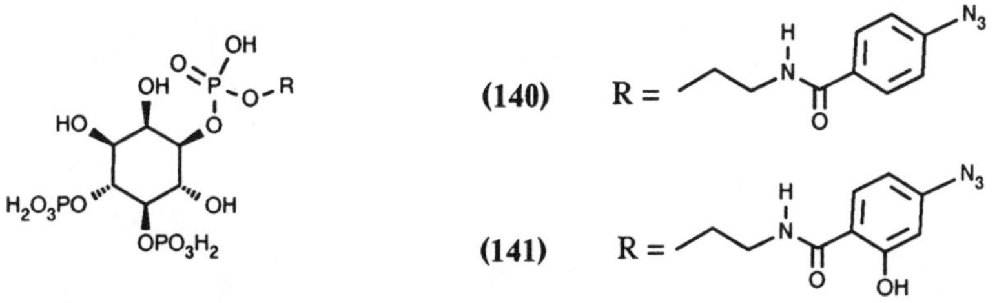

Fig 8 - 35

both of which bind to high affinity 1,4,5-IP$_3$ binding sites, with K_d *ca* 10-fold weaker than 1,4,5-IP$_3$ itself.[44] The analogue (140) was shown to release calcium from intracellular stores with a potency of *ca* one-tenth of that observed for 1,4,5-IP$_3$. A radioiodinated derivative of (140) was successfully used to photoaffinity label three 1,4,5-IP$_3$ binding proteins in rat pancreatic acinar cells.[44]

By taking advantage of the nucleophilic nature of phosphorothioates 1S,4,5-IP$_3$, *rac*-(108) see *Fig 8-27*, has been coupled directly with the fluorescent derivative (142) to give the fluorescently labelled analogue (143), *Fig 8-36*.[38] The labelled molecule (143) exhibited the expected fluorescence at 540 nm when excited at 460 nm, and was potent at releasing

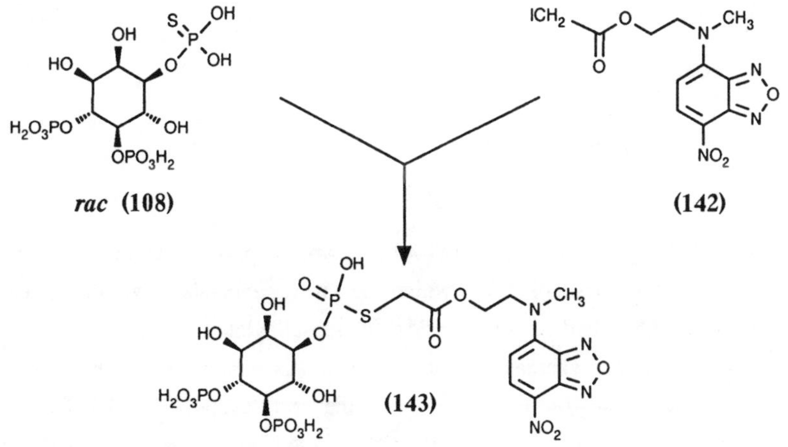

rac (108)

(142)

(143)

Fig 8 - 36

calcium from permeabilised cells, confirming that it is recognised by the 1,4,5-IP$_3$ receptor.
The esterified 1,4,5-IP$_3$ derivative (148) has been used in the synthesis of both a potential
radioiodinated photoaffinity label, and an affinity resin for the purification of 1,4,5-IP$_3$
binding proteins.[45] Treatment of the known protected inositol (144) with the phosphite
reagent (145) gave, after oxidation to the phosphate level, the fully protected compound

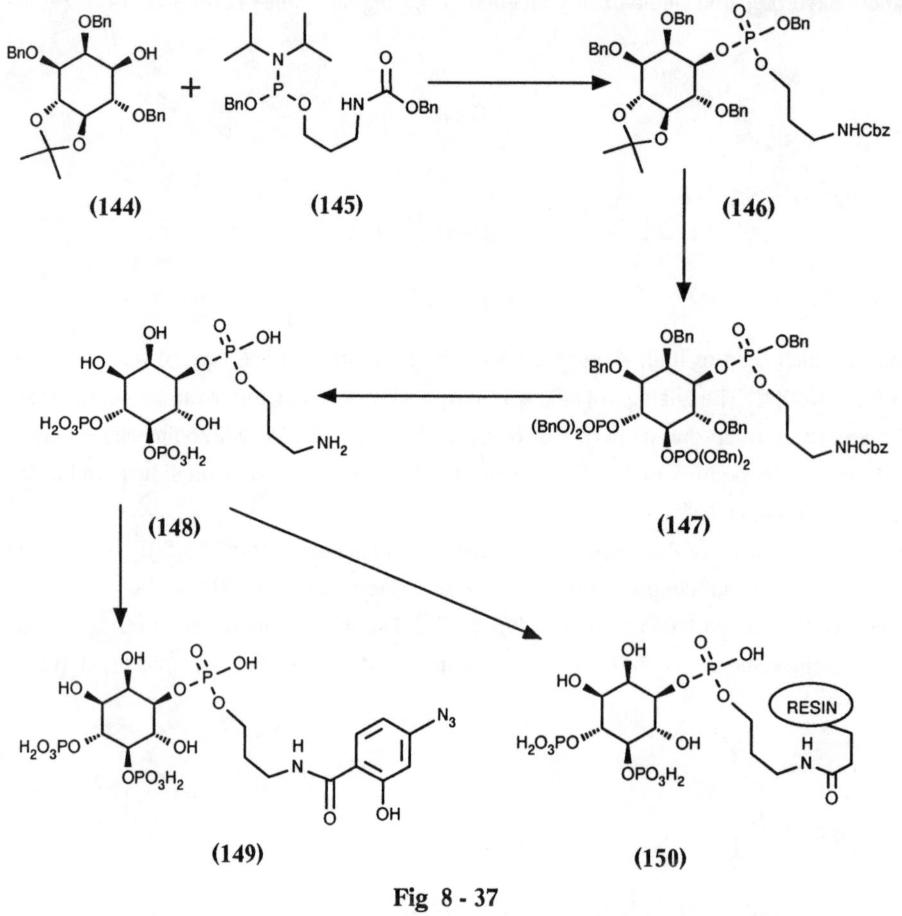

Fig 8 - 37

(146), *Fig 8-37*. Hydrolysis of the ketal in (146) followed by phosphorylation of the resulting
diol gave the protected trisphosphate, which on hydrogenolysis gave the ester (148).
Treatment of (148) with N-hydroxysuccinimido 4-azidosalicylate then gave the photoaffinity
labelled compound (149), whereas reaction with a suspension of a resin-immobilised
N-hydroxysuccinimide ester (Affi-Gel 10) gave the resin-supported 1,4,5-IP$_3$ derivative
(150). The incorporation of an ^{125}I label into (149) by substitution on the aromatic ring to
produce a radioiodinated photoaffinity label should be possible. Both (148) and (149)

displaced radiolabelled 1,4,5-IP$_3$ from receptor sites in brain, and both compounds were able to evoke calcium release from reconstituted liposome receptor preparations. In contrast, neither of the derivatives showed any interaction with the 5-phosphatase and 3-kinase enzymes, being neither substrates for, nor inhibitors of the enzymes. The immobilised analogue (150) has been successfully used for purification of 1,4,5-IP$_3$ receptors by affinity chromatography.

Another application of 1,4,5-IP$_3$ analogues is in the preparation of "caged" molecules. A caged molecule is an inert photolabile precursor of a biologically active substance, which can be used to introduce the substance into a biological system under controlled conditions in an inert form, and then release the active compound on photolysis.[46] A large number of caged molecules including for example caged ATP, GTP and cAMP have been prepared by the use of photolabile 2-nitrophenyl esters of phosphate groups, and have applicability to a range of biochemical problems. Esterification of 1,4,5-IP$_3$ with 1-(2-nitrophenyl)diazoethane gives a mixture of mono and multiply esterified materials.[47] Separation of the monoesterified

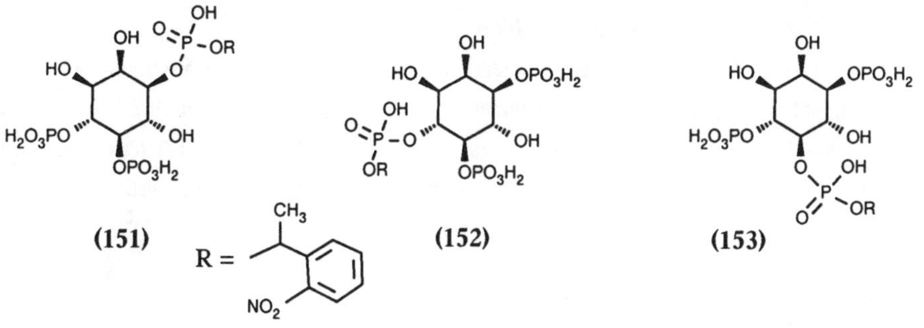

Fig 8 - 38

materials by HPLC and detailed structural analysis by 500 MHz proton NMR allowed the isomeric materials to be assigned as 1-, 4- and 5- monoesters, *ie* (151), (152) and (153) respectively, *Fig 8-38*. All three compounds released 1,4,5-IP$_3$ on irradiation. The 1-monoester (151) was a potent calcium-mobilising agent in its own right, and was also a substrate for the 5-phosphatase. Therefore, although fitting with previously established structure-activity results, (151) was not suitable as an inert caged 1,4,5-IP$_3$. In sharp contrast, both of the other isomers (152) and (153) were inactive in calcium release studies, even at high concentrations, and were not substrates for either the 5-phosphatase or the 3-kinase. As both of these molecules release 1,4,5-IP$_3$ on irradiation, with rate constants of the order of 300 s^{-1} , and in good quantum yield, they should find application in kinetic studies of 1,4,5-IP$_3$-mediated calcium mobilisation. The same authors reported preliminary results on this approach using a mixture of uncharacterised caged 1,4,5-IP$_3$ analogues.[48] Non-specific coupling of 1,4,5-IP$_3$ to *p*-azidobenzoic acid, to give caged analogues in *ca* 10% yield has also been reported.[49]

8.5.4 Phosphorus Replacement

The synthesis of the trissulphate (154) and trissulphamate (155) analogues of 1,4,5-IP$_3$ has been reported,[50] by direct functionalisation of the known racemic triol (95), *Fig 8-39*. No biological activity has been reported for these compounds.

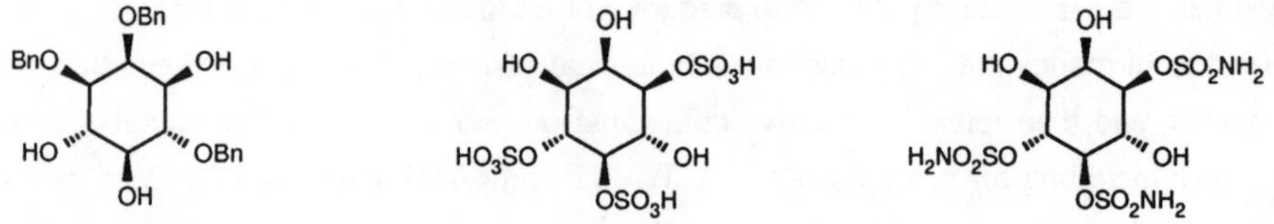

rac (95)   (154)   (155)

Fig 8 - 39

8.5.5 Inositol Modifications at the 2-Position

The first 2- position analogues reported were the mono and difluoro analogues of inositol 1,3,4-trisphosphate (1,3,4-IP$_3$).[51] Reaction of the protected alcohol (156) with diethylaminosulphur trifluoride (DAST) gave the equatorial fluoride (157) as expected, *Fig 8-40* (DAST normally introduces F atoms in place of OH groups with inversion of

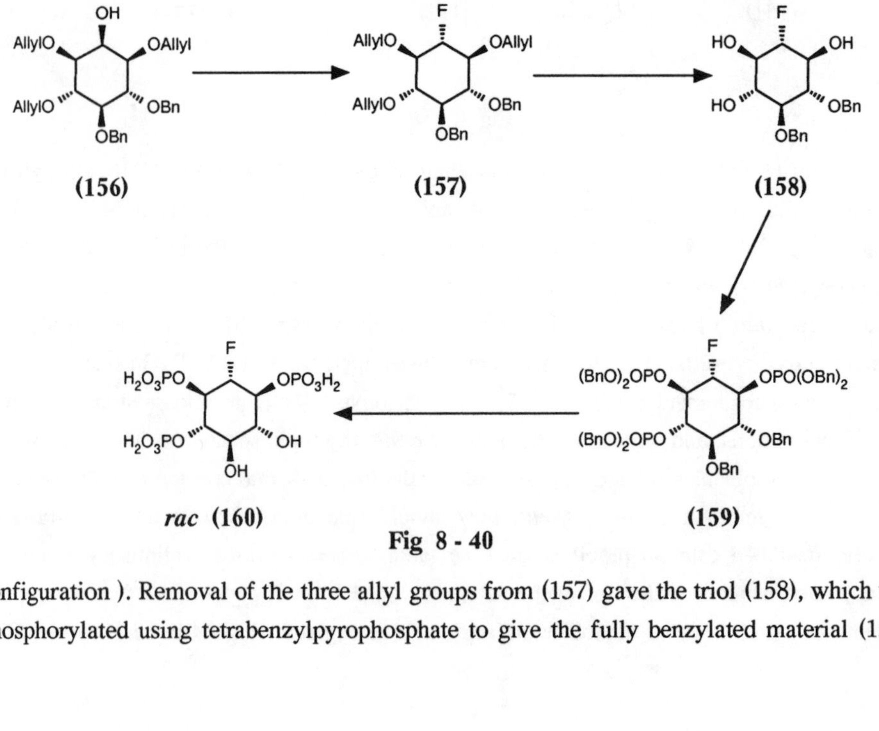

(156)   (157)   (158)

rac (160)   (159)

Fig 8 - 40

configuration). Removal of the three allyl groups from (157) gave the triol (158), which was phosphorylated using tetrabenzylpyrophosphate to give the fully benzylated material (159).

Hydrogenolysis of (159) then gave the desired analogue 2-fluoro-2-deoxy-1,3,4-IP$_3$, *rac-(160)*. Oxidation of the alcohol (156) to the ketone (161), followed by treatment with DAST for a prolonged period gave the difluoro compound (162), *Fig 8-41*. Removal of the three allyl groups from (162), followed by phosphorylation and deprotection as above then gave the difluoro analogue 2,2-difluoro-2-deoxy-1,3,4,-IP$_3$, *rac*-(163). The authors reported no biological activity for these compounds.

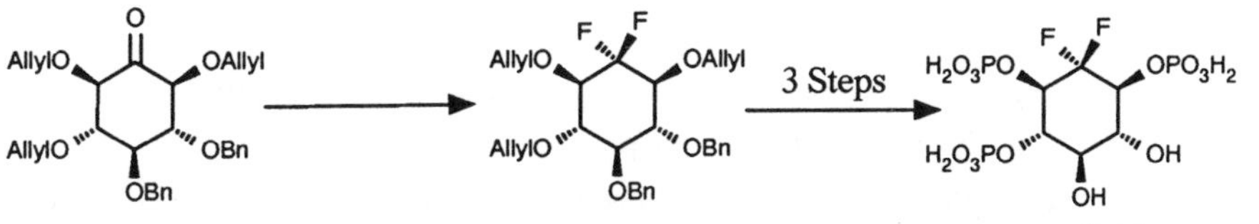

(161)   **(162)**   *rac* **(163)**

Fig 8 - 41

Following a similar approach, the same authors reported the synthesis of the corresponding mono and difluoro analogues of 1,4,5-IP$_3$.[52] Treatment of the known alcohol (164) with DAST gave the fluoride (165), *Fig 8-42*. Removal of the allyl and acetal groups from (165), followed by phosphorylation and deprotection then gave 2-fluoro-2-deoxy-1,4,5-IP$_3$, *rac-(166)*. Oxidation of (164) gave the ketone (167), which on treatment with DAST and

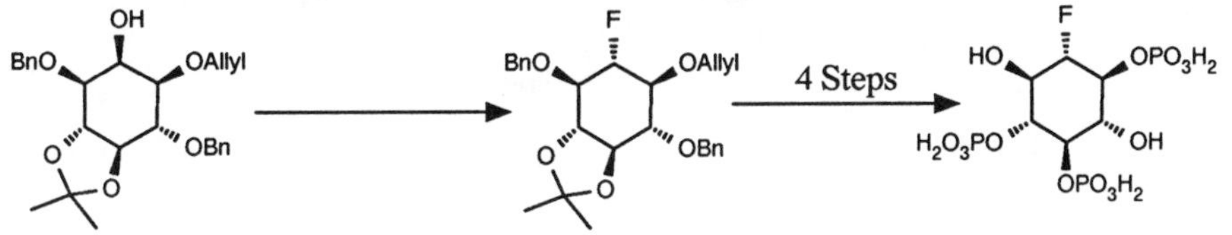

(164)   **(165)**   *rac* **(166)**

Fig 8 - 42

subsequent removal of the allyl and acetal protecting groups gave the triol (168), *Fig 8-43*. Phosphorylation of (168) followed by deprotection then gave the difluoro analogue 2,2-difluoro-2-deoxy-1,4,5-IP$_3$, *rac*-(169). The authors reported that both of these molecules, (166) and (169), showed high affinity for the rat brain 1,4,5-IP$_3$ receptor, and that both were agonists able to release calcium from permeabilised cells. A detailed biological evaluation of these molecules has subsequently appeared,[53] indicating that both are full agonists, only slightly lower in potency than 1,4,5-IP$_3$ itself. Both molecules released calcium with similar kinetic profiles to 1,4,5-IP$_3$, and both were substrates for the 3-kinase. The monofluoro compound (166) was a weak substrate for the 5-phosphatase, whereas the difluoro analogue (169) was a potent inhibitor of 5-phosphatase activity, $K_i = 26$ μM, and was not hydrolysed by the enzyme. The authors concluded that the 2- hydroxyl group did not play an absolute

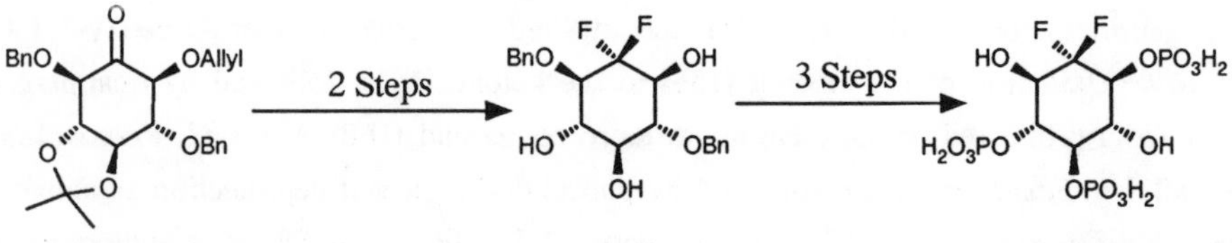

Fig 8 - 43

role in either receptor binding or agonist activity in these molecules, although it could have an important recognition role in the enzymic processing of 1,4,5-IP$_3$.

These results are supported by the biological activity of a series of 2-substituted analogues of 1,4,5-IP$_3$, whose syntheses have not been described in full to date.[54,55,56] The general route to these analogues involves reaction of the known selectively silylated inositol (170), see *Fig 6-8*, or its enantiomer with various alkylating groups followed by desilylation, phosphorylation, and deprotection to give the 2-substituted analogues (171)-(175), *Fig 8-44*.

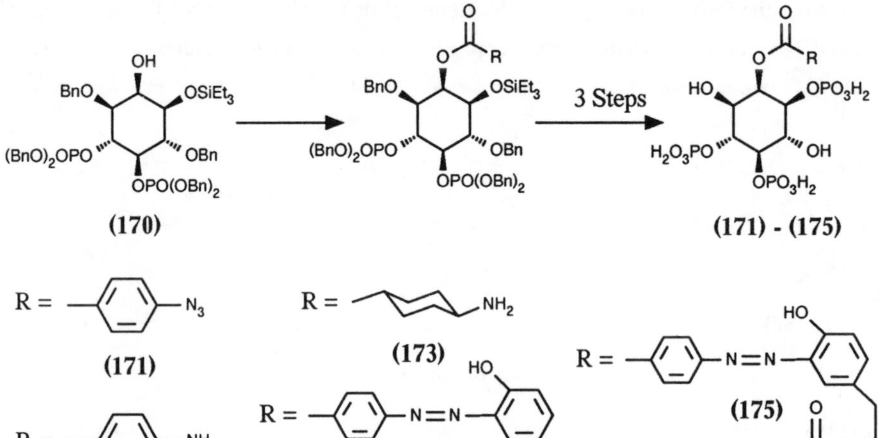

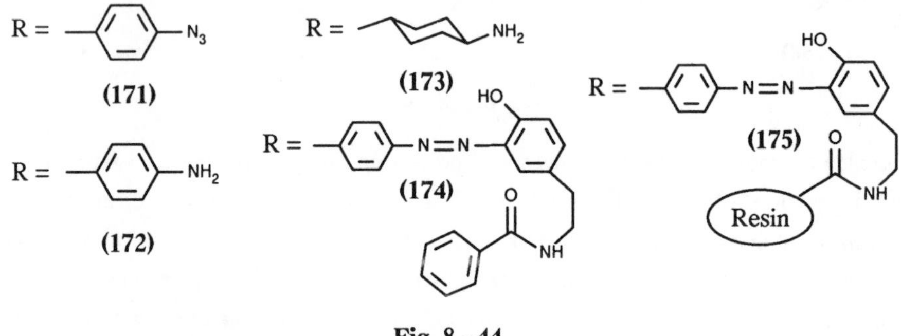

Fig 8 - 44

These compounds provided some very interesting insight into the recognition of 1,4,5-IP$_3$ by its receptor and metabolic enzymes. All of the racemic compounds (171)-(174) were full agonists at the 1,4,5-IP$_3$ receptor, releasing calcium with between one-half and one-fiftieth the potency of 1,4,5-IP$_3$, and all competed with 1,4,5-IP$_3$ for the receptor binding site.[54] In addition, all of the racemic compounds (171)-(174) inhibited the 5-phosphatase with K_i

values below that observed for 1,4,5-IP$_3$, whereas although they also inhibited the 3-kinase, K_i values were some 2 - 50 times higher than for 1,4,5-IP$_3$. The authors also reported that 2-deoxy-1,4,5-IP$_3$ showed a similar profile to these analogues. Resolution of the two amino analogues (172) and (173) and subsequent testing showed that, as predicted, the D-isomers (*ie* those having the same configuration as naturally occurring D-1,4,5-IP$_3$) were more active in binding to 1,4,5-IP$_3$ receptor sites than the L-isomers.[55] The D-isomers were also more active as inhibitors of the 3-kinase than their L-enantiomers. In stark contrast to these results, the D-isomers showed no measurable inhibition of the 5-phosphatase, but were slowly hydrolysed by the enzyme. The L-isomers however were potent inhibitors of 5-phosphatase activity, and were not hydrolysed by the enzyme. This switch of isomer specificity may reflect the inherent symmetry of the inositol system, and the potent inhibitor / weak substrate isomer separation is very reminiscent of the observations made earlier on inositol monophosphatase, Section 8.3.3.

This approach has been extended to prepare 1,4,5-IP$_3$ affinity columns by linking analogues such as (175) to resin supports and these resins have been applied to the purification of 1,4,5-IP$_3$ binding proteins, and of the 5-phosphatase and 3-kinase enzymes.[56] Finally these authors have prepared (176), *Fig 8-45*, which combines a photosensitive linkage with a

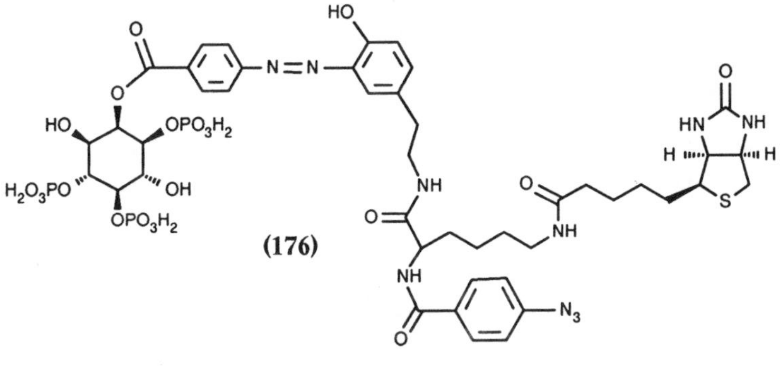

Fig 8 - 45

biotin-avidin complexing moiety, to produce a photolabile derivative, detectable by non-radiochemical means.[57] The resulting analogue (176) displays almost equal inhibition of 5-phosphatase activity as 1,4,5-IP$_3$ itself, and on photolysis successfully labelled the enzyme. From these results it is clear that the 1,4,5-IP$_3$ receptor and the 5-phosphatase and 3-kinase can all tolerate very bulky groups at the 2-position of their target 1,4,5-IP$_3$,and that the presence of a 2-hydroxyl group is not an absolute requirement. The exact requirements of the 3-kinase and 5-phosphatase for substrate recognition are probably quite different, and the 2-hydroxyl group may play a role here.

8.5.6 Inositol Modifications at the 3-Position

The first reported analogue of 1,4,5-IP$_3$ modified at the 3-position was the inverted hydroxyl derivative *rac*-(181)[58] whose correct systematic name is *chiro*-inositol 2,3,5-trisphosphate, as inversion of the 3-hydroxyl group of *myo*-inositol gives *chiro*-inositol, see Chapter 1. Benzylation of the known diol (177), see *Fig 6-16*, followed by *cis*-hydroxylation of the double bond gave the diol (178), *Fig 8-46*. Mono-protection of the diol (178) gave

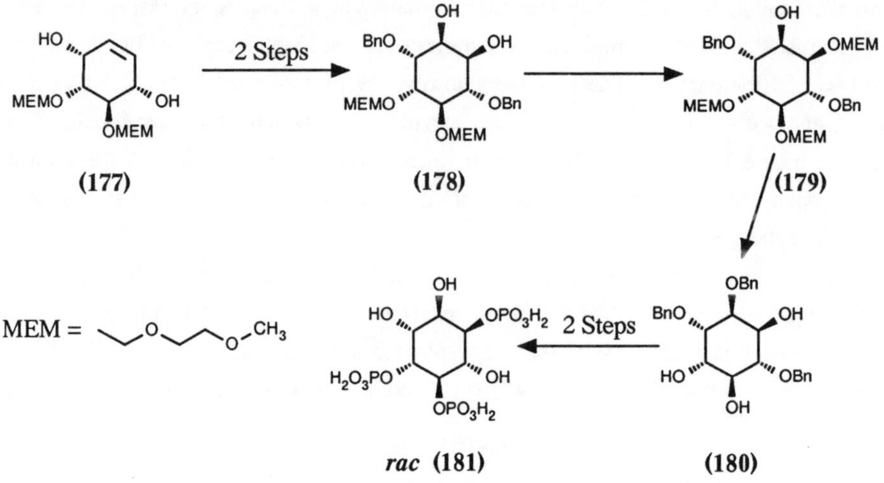

Fig 8 - 46

preferentially the equatorial derivative (179). Benzylation of (179) followed by removal of the MEM groups then gave the triol (180), which on phosphorylation and deprotection gave racemic *chiro*-inositol 2,3,5-trisphosphate, *rac*-(181). The authors gave no biological data for this racemic material.

A synthesis of L-*chiro*-inositol 2,3,5-trisphosphate which uses L-quebrachitol (182) as starting material has subsequently appeared.[59] Note that L-*chiro*-inositol 2,3,5-trisphosphate has the same absolute configuration as D-*myo*-inositol 1,4,5-trisphosphate, with the 3-position inverted. Demethylation of (182) gave free L-*chiro*-inositol (183), *Fig 8-47*. Treatment of (183) with excess tributyltin oxide, followed by excess benzyl chloride gave the triol (184) as a major product. Benzoylation of (184) then gave the fully protected *chiro*-inositol (185) which on catalytic hydrogenation gave the triol (186). Phosphitylation of (186), followed by oxidation to the protected trisphosphate and subsequent deprotection then gave the desired L-*chiro*-inositol 2,3,5-trisphosphate, L-(187).

The authors reported[59] that L-(187) is a potent agonist at the 1,4,5-IP$_3$ receptor, releasing calcium with *ca* one-fifth the potency of 1,4,5-IP$_3$. As hoped, L-(187) was not a substrate for the 3-kinase, and produced long-lived calcium mobilisation. Unexpectedly however, L-(187)

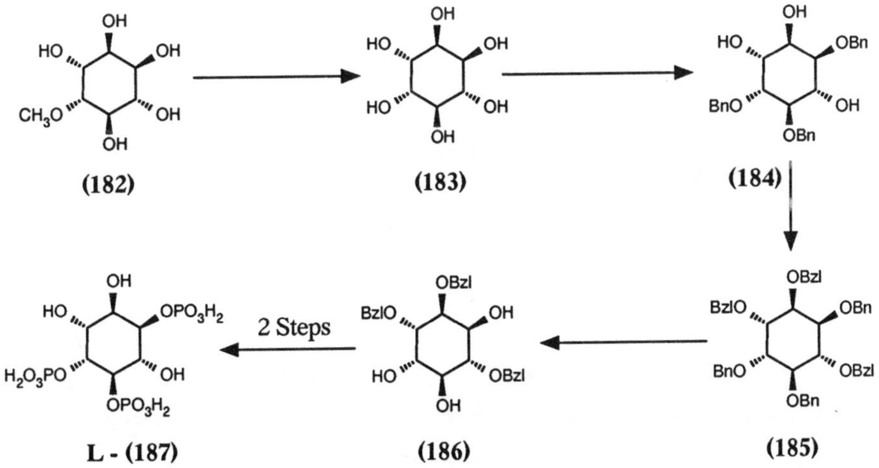

Fig 8 - 47

was not a substrate for the 5-phosphatase either, raising the question of how inversion of a hydroxyl group in this relatively remote position destroys the substrate activity of L-(187) for phosphate hydrolysis at the 5-position.

L-Quebrachitol (182) has also been used as a starting material for the synthesis of optically pure D-3-deoxy-*myo*-inositol 1,4,5-IP$_3$, D-(192)[60] and D-3-deoxy-3-fluoro-*myo*-inositol 1,4,5-IP$_3$, D-(195).[61] Conversion of (182) to its bisketal, followed by Barton deoxygenation gave the fully protected viburinol derivative (188), *Fig 8-48*.[60] Treatment of (188) with BBr$_3$

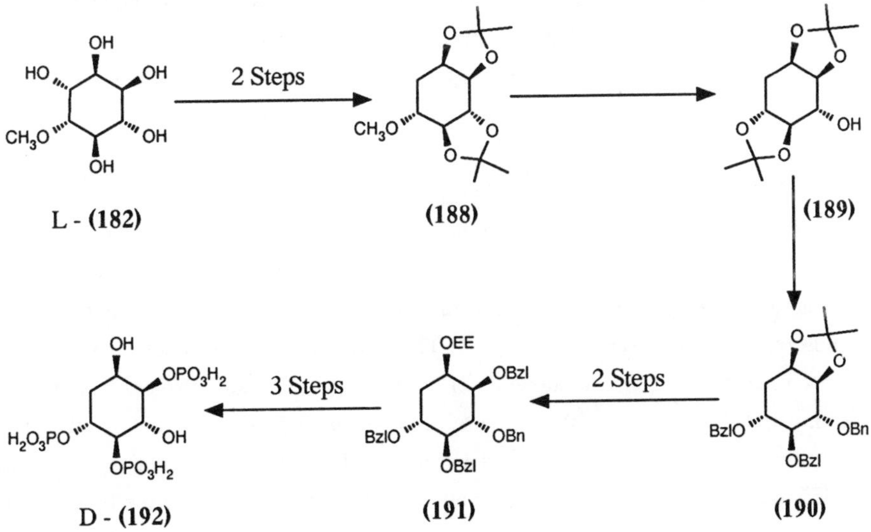

Fig 8 - 48

removed all of the protecting groups, and reprotection as the bisketal gave a 1:1.3 mixture of bisketal (189) and a regioisomeric bisketal. Benzylation of the free hydroxyl group in (189), followed by selective hydrolysis of the less stable *trans* ketal and benzoylation of the resulting diol gave the fully protected compound (190). Cleavage of the remaining ketal group, followed by selective benzoylation of the equatorial hydroxyl group, and protection of the remaining axial hydroxyl group as the ethoxyethyl ether then gave the fully protected derivative (191). Finally cleavage of the three benzoyl groups, phosphorylation of the resulting triol, and deprotection gave the desired product, D-3-deoxy-*myo*-inositol 1,4,5-IP$_3$, D-(192). The authors reported that D-(192) acted as a full agonist at the 1,4,5-IP$_3$ receptor, releasing calcium with a profile similar to 1,4,5-IP$_3$. The molecule is presumably, by definition, not a substrate for the 3-kinase, but no data on the interaction of D-(192) with either the 3-kinase or the 5-phosphatase has been reported. By following essentially the same

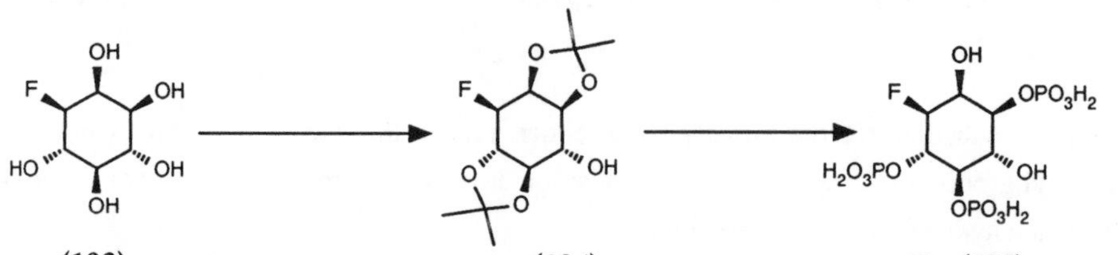

Fig 8 - 49

route, 3-deoxy-fluoro-*myo*-inositol (193)[62] was converted *via* the bisketal (194), *Fig 8-49,* (equivalent to the bisketal (189) in *Fig 8-48*) into D-3-deoxy-3-fluoro-*myo*-inositol 1,4,5-IP$_3$, D-(195).[61] The authors reported that D-(195) acted as a full agonist at the 1,4,5-IP$_3$ receptor, releasing calcium with a profile similar to 1,4,5-IP$_3$, but again no data on the interaction of D-(195) with either the 3-kinase or the 5-phosphatase has been reported.

8.5.7 Inositol Modifications at the 6-Position

Using the chemistry developed in their synthesis of 1,4,5-IP$_3$, Ley's group have reported the synthesis of a series of analogues modified at the 6-position.[63,64,65] All of the derivatives were prepared by opening of the epoxide function in the protected compound (196), *Fig 8-50*, whose synthesis is described in *Fig 6-9*. Thus reduction of (196) with LiAlH$_4$ gave the deoxy intermediate (197), treatment of (196) with tris(diethylamino)sulphonium difluorotrimethylsilicate gave the fluoro compound (198), reaction with lithium dimethyl(cyano)copper (I) gave the methylated intermediate (199), and reaction with sodium methoxide gave the methoxy compound (200). Hydrogenation to remove the benzyl groups was followed by phosphorylation and final deprotection to give the 1,4,5-IP$_3$ analogues (201)-(204). The authors reported no biological data for these compounds.

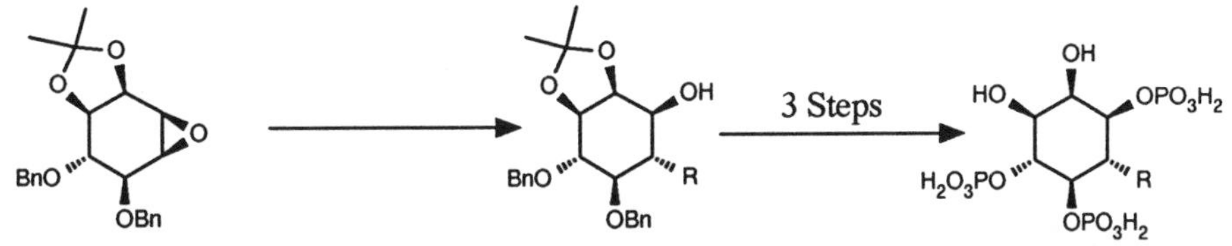

(196)

(Synthesis see *Fig 6-9*)

(197) R = H
(198) R = F
(199) R = CH_3
(200) R = OCH_3

(201) R = H
(202) R = F
(203) R = CH_3
(204) R = OCH_3

Fig 8 - 50

Reports concerning the biological activity of both 6-deoxy 1,4,5-IP_3 (201),[66] and 6-methoxy 1,4,5-IP_3 (204)[67] have subsequently appeared, but no details of the syntheses of these molecules were given. 6-Deoxy 1,4,5-IP_3, (201) is a full agonist at the 1,4,5-IP_3 receptor, mobilising calcium with a potency *ca* 70-fold weaker than natural 1,4,5-IP_3.[66] Analogue (201) inhibited 5-phosphatase activity potently and 3-kinase activity less potently, and appeared to be phosphorylated by the 3-kinase. In contrast, (201) is not a substrate for the 5-phosphatase. 6-Methoxy 1,4,5-IP_3, (204) is also a full agonist at the 1,4,5-IP_3 receptor, with potency *ca* 100-fold weaker than 1,4,5-IP_3.[67] Analogue (204) was also a potent inhibitor of 5-phosphatase activity, but was not hydrolysed by the enzyme. Inhibition of the 3-kinase was also observed, but with weaker affinity.

8.5.8 Multiple Modifications

I have attempted to group here compounds which can be considered to be derived from 1,4,5-IP_3 by more than one modification of the parent molecule. This definition is of course very subjective, as cyclohexanol can be considered to be derived from 1,4,5-IP_3 by deletion of two hydroxyl groups and three phosphate groups ! Detailed comparisons between compounds studied by different groups are impossible due to the different systems used to assay calcium release and the different formats used to express biological activity. For this reason the compounds are grouped in sets studied under the same conditions, and thus each group forms an internally consistent, but necessarily somewhat isolated, series.

The first study of this type reported the activity of the compounds shown in *Fig 8-51.* The biological activity of these compounds was determined in bovine aorta derived tissues, and is summarised in *Table 8-1* in terms of calcium release (EC_{50}), inhibition of the 3-kinase (IC_{50}), inhibition of the 5-phosphatase (K_i) and hydrolysis by the 5-phosphatase (observable phosphate release under standard conditions), with the compounds arranged in order of descending potency in the calcium release assay.

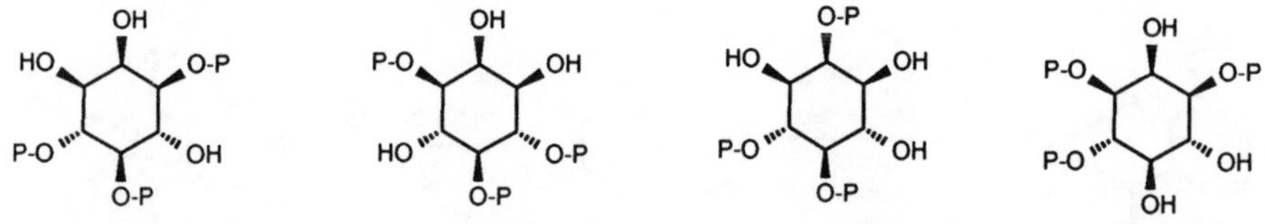

D -(205)   L -(206)   DL -(207)   DL -(208)

P = OPO₃H₂

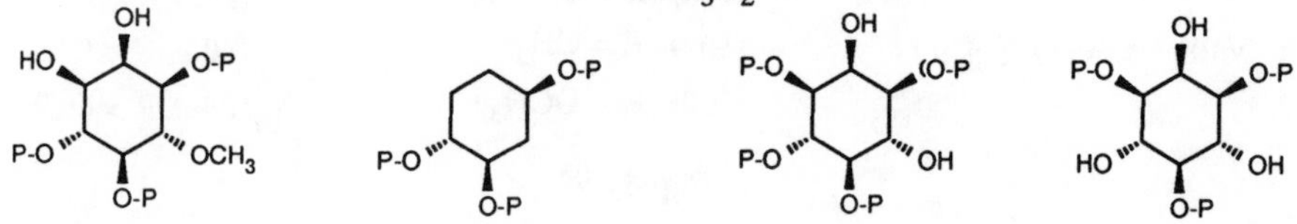

DL -(204)   DL -(211)   DL -(210)   (209)

Fig 8 - 51

Table 8 - 1 . Biological activity of multiply modified compounds

Compound	EC$_{50}$ Ca^{++} (μM)	IC$_{50}$ kinase (μM)	IC$_{50}$ phosphatase (μM)	Substrate 5-phosphatase
(205)	0.3	2.6	12	Yes
(207)	8.7	1868	131	Yes
(211)	40	327	137	No
(210)	65	90	1.4	Yes
(204)	65	319	52	No
(208)	120	1847	32	No
(206)	220	292	39	No
(209)	630	1687	45	No

All of the compounds in *Fig 8-51* released calcium from permeabilised smooth muscle cells, and all were full agonists. The most potent compound in this series was D-1,4,5-IP₃ (205), with DL-2,4,5-IP₃ (207) being only some 20-fold less potent. Removal of all of the hydroxyl groups to give (211) resulted in a *ca* 100-fold potency loss, as did methylation at the 6-position (204) or addition of an additional phosphate group at the 3-position (210) . Removal of the vicinal phosphate groups as in (209) gave a further *ca* 10-fold drop in potency.

The most potent compound after D-1,4,5-IP$_3$ (205) in the inhibition of the 3-kinase was DL-1,3,4,5-IP$_4$ (210), with a *ca* 50-fold drop in potency. The enantiomeric L-1,4,5-IP$_3$ (206), 6-methoxy (204) and fully dehydroxylated compound (211) were all *ca* 100-150 times less potent than natural (205), *ie* the enzyme's normal substrate.

The 5-phosphatase showed higher affinity for DL-1,3,4,5-IP$_4$ (210), than for D-1,4,5-IP$_3$ (205), as previously reported[68], and surprisingly L-1,4,5-IP$_3$ (206) was only some 3-fold less potent than its enantiomer. The enzyme showed an interesting lack of selectivity, with (204), (208) and (209) all having similar activity to (206).

In contrast, the enzyme only hydrolysed those analogues having a 5-phosphate vicinal to a free phosphate at the 4-position and a free hydroxyl group at the 6-position, *ie* (205), (207) and (210). As neither the enantiomeric 1,4,5-IP$_3$ (206) nor the 6-methoxy compound (204) were substrates the enzyme seems to require both a free hydroxyl at 6- and the D-configuration for the 4,5-vicinal phosphate groups for substrate activity.

The release of calcium from vacuoles of *Neurospora crassa* involves a putative receptor protein, which is clearly different from, and significantly less specific than the bovine receptor, as all of the compounds (212) - (215) shown in *Fig 8-52* have been reported to release calcium in this system,[69] with potency almost equal to D-1,4,5-IP$_3$ (205).

D-2,4,5-IP$_3$, (216) has been shown to be a full agonist in several calcium release studies, with

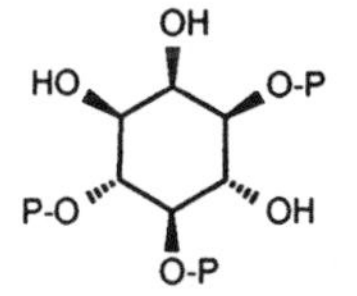

D -(205)

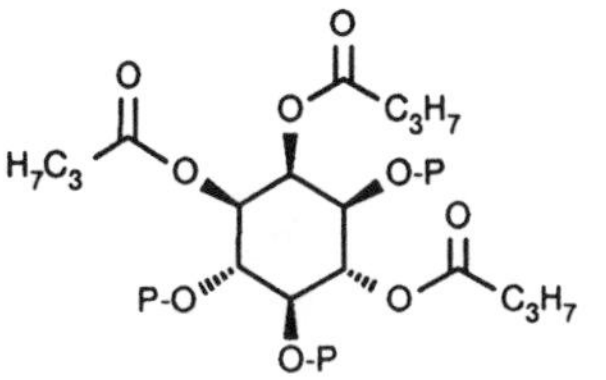

(212)

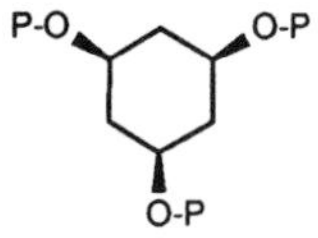

(213)

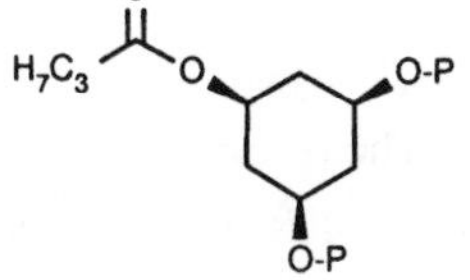

$P = PO_3H_2$

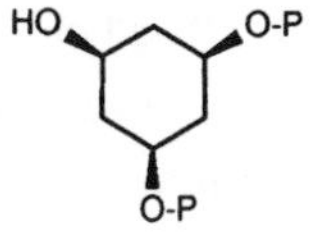

(214)

Fig 8 - 52

(215)

a potency *ca* 2-fold higher than that observed for the racemic material, *ie* (207).[70,71,72]

Inversion of the hydroxyl group at the C-1 position of D-2,4,5-IP$_3$ (216) to give

D-*chiro*-inositol 1,3,4-IP$_3$ (217) results in no loss of potency[72] in calcium release, *Fig 8-53*. The results for a series of analogues are given in *Table 8-2*, and it can be seen that the enantiomeric L-trisphosphates (218) and (219) were *ca* 30-fold less potent than their D-counterparts in this study. The methylene phosphate (220) is of very similar activity to the trisphosphate from which it may be considered to be derived, *ie* (211), *Fig 8-51* and *Table 8-1*. The importance of the vicinal 4,5- bisphosphate grouping for agonist activity at the 1,4,5-IP$_3$ receptor is underlined by the activity of the simple cyclohexane bisphosphate (221), and the somewhat lower activity of *rac*-4,5-IP$_2$ (222).

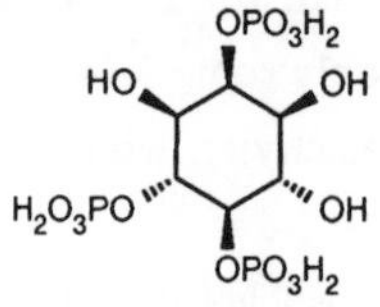

(216) 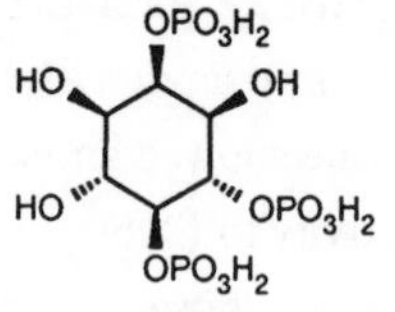   (218) 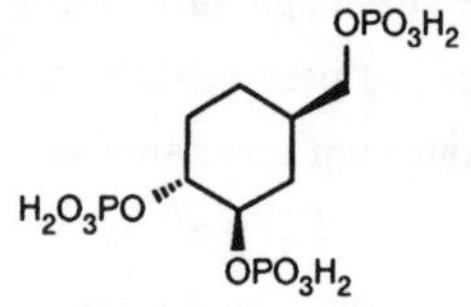   (220)

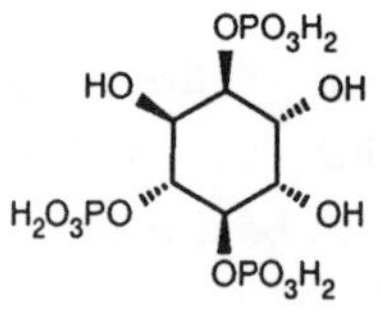

(217) 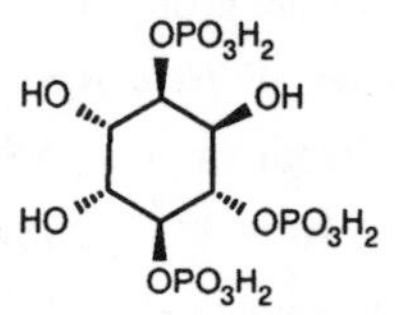   (219) 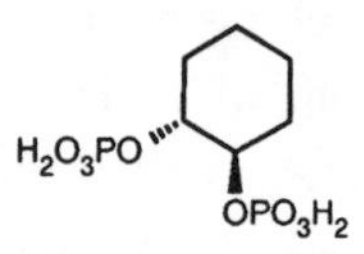   (221)

Fig 8 - 53

Table 8 - 2 . Biological activity of analogues of 1,4,5-IP$_3$

Compound	(205)	(216)	(217)	(221)	(220)	(222)	(218)	(219)
EC$_{50}$ Ca^{++} (μM)	0.13	4	4	22	30	70	105	125

The synthesis and biological activity of a series of analogues of 1,4,5-IP$_3$ has been reported in two posters presented by the Fisons medicinal chemistry group.[73] The results of these studies are summarised in *Table 8-3* ,for the analogues shown in *Fig 8-54*. The authors concluded that the minimum requirement for recognition at the IP$_3$ receptor, and calcium release was the all-*trans* cyclohexane-1,3/2-triol 1,2-bisphosphate moiety, and that only compounds having vicinal bisphosphates showed any activity. This contrasts with the activity reported for the simple bisphosphate (221) by Denis and Ballou.[72] Other derivatives not shown in *Fig*

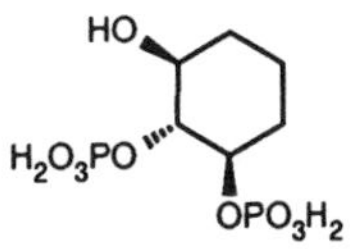

(222) 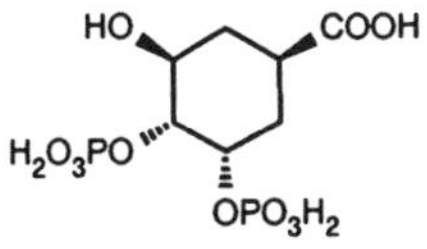 (228) 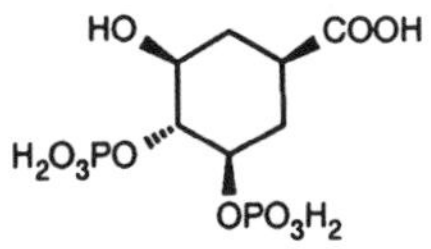 (223) 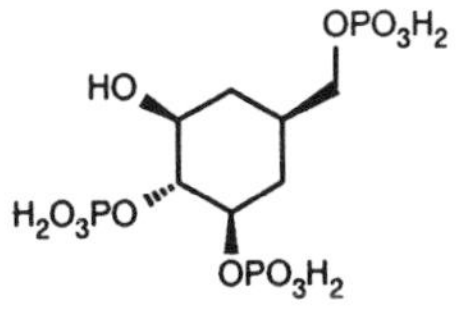 (227)

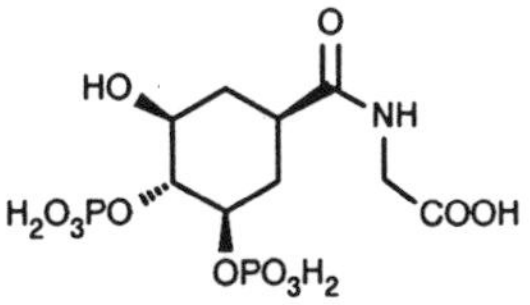

(226) 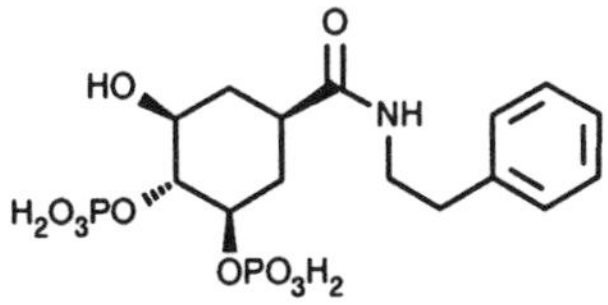 (225) 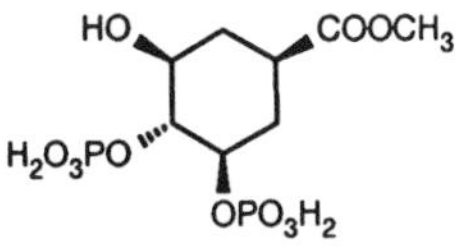 (224)

Fig 8 - 54

Table 8 - 3 . Biological activity of analogues of 1,4,5-IP$_3$

Compound	EC$_{50}$ IP$_3$ receptor (μM)	EC$_{50}$ Ca^{++} (μM)	IC$_{50}$ phosphatase (μM)
(221)	>300	Inactive	1000
(222)	45	400	>1000
(223)	5	100	130
(224)	12	150	1000
(225)	10	150	>300
(226)	3	40	300
(227)	1.5	20	100
(228)	>100	Inactive	50

8-54 indicated that any modification of the phosphate groups, *eg* by esterification, led to loss of both IP$_3$ binding and 5-phosphatase inhibitory activity. In addition a wide range of substituents could be accommodated at the 1-position of the analogues without loss of IP$_3$ receptor binding or calcium mobilising activity, whilst generally a third acidic group was required at this position for inhibition of the 5-phosphatase. Compounds with *cis*-vicinal bisphosphate groups seemed to be at least as active as their *trans* counterparts for inhibition of the 5-phosphatase, compare (223) and (228), but much less potent in IP_3 binding. None of the compounds showed any evidence of antagonist activity at the IP_3 receptor. The authors concluded that the structural requirements for agonist activity at the IP_3 receptor, and for inhibitory activity against the 5-phosphatase were quite different.

8.6 Analogues of Inositol Tetrakisphosphates

Very few analogues of inositol tetrakisphosphates have been reported to date. Using the same methodology as for the trisphosphates, *Fig 8-26*,[25] Dreef *et al* have prepared the 5-phosphorothioate, 1,3,4,5S-IP_4 (230) from the alcohol (229), *Fig 8-55*. Preliminary results indicated that (230) is a competitive inhibitor of the 3-kinase.

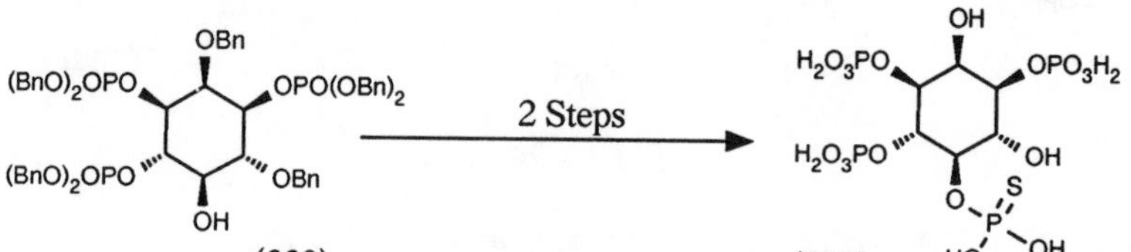

Fig 8 - 55

The diol (231) has been used for the preparation of both an affinity probe, and an affinity resin based on 1,3,4,5-IP_4 esterified at the 1-phosphate.[74] Selective protection of the equatorial alcohol at position 1 of (231) as the methoxymethyl ether, followed by benzylation of the remaining free hydroxyl group, and removal of the methoxymethyl group gave the alcohol (232), *Fig 8-56*. Phosphitylation of (232) using a substituted phosphine, followed by oxidation to the protected phosphate and subsequent removal of the allyl groups then gave the protected triol (233). Phosphorylation of (233), and subsequent removal of all of the benzyl protecting groups then gave the free tetrakisphosphate ester (234). The ester (234) showed an affinity for the rat brain 1,3,4,5-IP_4 receptor of *ca* one-tenth that of 1,3,4,5-IP_4 itself. Two probes were prepared from (234), both by direct reaction of (234) with N-hydroxysuccinimidyl groups, giving the photoaffinity probe (235) and the affinity resin (236). The resin (236) effectively bound all of the IP_4 binding affinity in a partially purified solubilised cerebellar membrane protein preparation, and provides a new tool for the isolation and study of IP_4 binding proteins.

From these initial studies it seems that in common with the 1,4,5-IP_3 receptor, the 1,3,4,5-IP_4 receptor tolerates modification of the phosphate group at the 1- position with a reduction in observed potency of *ca* 10-fold. General aspects of the recognition of the inositol phosphates by proteins were reviewed in 1989.[75]

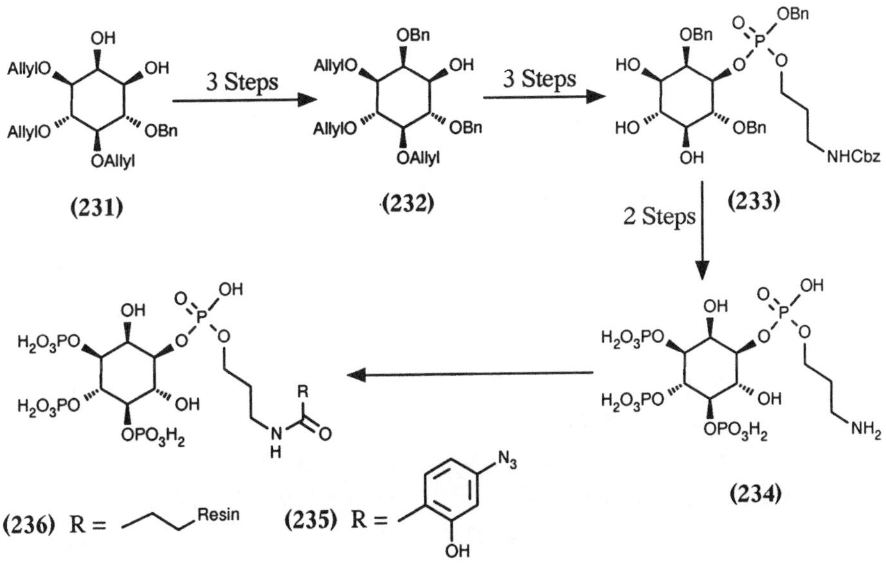

Fig 8 - 56

References

1) T.Metschies, C.Schultz and B.Jastorff, *Tetrahedron Lett.*, 1988, **29**, 3921.

2) G.R.Baker, D.C.Billington and D.Gani, *Bioorg.Med.Chem.Lett.*, 1991, **1**, 17.

3) G.R.Baker, D.C.Billington and D.Gani, *Tetrahedron*, 1991, **47**, 3895.

4) D.C.Billington, R.Baker, I.M.Mawer and J.J.Kulagowski, *J.Chem.Soc.Chem. Commun.*, 1987, 314.

5) A.M.Cooke, R.Gigg and B.V.L.Potter, *J.Chem.Soc.Chem.Commun.*, 1987, 1525.

6) P.M.Burgers and F.Eckstein, *Tetrahedron Lett.*, 1978, 3835.

7) J.J.Kulagowski, *Tetrahedron Lett.*, 1989, **30**, 3869.

8) D.E.Kieley, G.J.Abruscato and V.Baburao, *Carbohydrate Res.*, 1974, **34**, 307.

9) J.J.Kulagowski, R.Baker and S.R.Fletcher, *J.Chem.Soc.Chem.Commun.*, 1991, 1649.

10) C.E.Dreef, M.Douwes, C.J.J.Elie, G.A. van der Marel and J.H. van Boom, *Synthesis*, 1991, 443.

11) R.Baker, J.J.Kulagowski, D.C.Billington, P.D.Leeson, I.C.Lennon and N.Liverton, *J.Chem.Soc.Chem.Commun.*, 1989, 1383.

12) R.Gigg and C.D.Warren, *J.Chem.Soc.(C)*, 1969, 2367.

13) R.Baker, P.D.Leeson, N.J.Liverton and J.J.Kulagowski, *J.Chem.Soc.Chem.Commun.*, 1990, 462.

14) G.H.Posner and D.Z.Rogers, *J.Am.Chem.Soc.*, 1977, **99**, 8208.

15)   R.Baker, C.Carrick, P.D.Leeson, I.C.Lennon and N.Liverton, *J.Chem.Soc.Chem. Commun.*, 1991, 298.

16)   D.P.Phillion and S.S.Andrew, *Tetrahedron Lett.*, 1986, 1477.

17)   M.R.Hamblin, J.S.Flora and B.V.L.Potter, *Biochem.J.*, 1987, **246**, 771.

18)   M.R.Hamblin, B.V.L.Potter and R.Gigg, *Biochem.Soc.Trans.*, 1987, **15**, 415.

19)   N.D.Sinha, J.Biernat, J.McManus and H.Koster, *Nucleic Acids Res.*, 1984, **12**, 4539.

20)   P.Westerduin, H.A.M.Willems and C.A.A. van Boekel, *Tetrahedron Lett.*, 1990, **31**, 6915.

21)   C.E.Dreef, R.J.Tuinman, A.W.M.Lefeber, C.J.J.Elie, G.A. van der Marel and J.H. van Boom, *Tetrahedron*, 1991, **47**, 4709.

22)   C.E.Dreef, C.J.J.Elie, P.Hoogerhout, G.A. van der Marel and J.H. van Boom, *Tetrahedron Lett.*, 1988, **29**, 6513.

23)   F.Eckstein and G.Gish, *Trends Biochem.Sci.*, 1989, **14**, 97.

24)   A.M.Cooke, B.V.L.Potter and R.Gigg, *Tetrahedron Lett.*, 1987, **28**, 2305.

25)   C.E.Dreef, G.W.Mayr, J-P.Jansze, H.C.P.F.Roelen, G.A. van der Marel and J.H. van Boom, *Bioorg.Med.Chem.Lett.*, 1991, **1**, 239.

26)   A.L.Willcocks, B.V.L.Potter, A.M.Cooke and S.R.Nahorski, *Eur.J.Pharmacol.*, 1988, **155**, 181.

27)   C W.Taylor, M.J.Berridge, A.M.Cooke and B.V.L.Potter, *Biochem.J.*, 1989, **259**, 645.

28)   C.W.Taylor, M.J.Berridge, A.M.Cooke and B.V.L.Potter, *Biochem.Biophys. Res.Commun.*, 1988, **150**, 626.

29)   J.Struptish, A.M.Cooke, B.V.L.Potter, R.Gigg and S.R.Nahorski, *Biochem.J.*, 1988, **253**, 901.

30)   F.Thevenod, M.Dehlinger-Kremer, T.P.Kemmer, A.L.Christian, B.V.L.Potter and I.Schulz, *J.Membr.Biol.*, 1989, **109**, 173.

31)   L.Changya, R.F.Irvine, D.V.Gallacher, B.V.L.Potter and O.H.Petersen, *J.Membr. Biol.*, 1989, **109**, 85.

32)   A.M.Cooke, S.R.Nahorski and B.V.L.Potter, *FEBS Lett.*, 1989, **242**, 373.

33)   S.T.Safrany, R.J.H.Wojcikiewicz, J.Strupish, J.McBain, A.M.Cooke, B.V.L.Potter and S.R.Nahorski, *Mol.Pharm.*, 1991, **39**, 754.

34)   R.J.H.Wojcikiewicz, A.M.Cooke, B.V.L.Potter and S.R.Nahorski, *Eur.J.Biochem.*, 1990, **192**, 459.

35)   A.M.Cooke, N.J.Noble, S.Payne, R.Gigg and B.V.L.Potter, *J.Chem.Soc.Chem. Commun.*, 1989, 269.

36)   J.Gigg, R.Gigg, S.Payne and R.Conant, *J.Chem.Soc.Perkin Trans. 1*, 1987, 423.

37)   A.M.Cooke, N.J.Noble, R.Gigg, A.L.Willcoks, J.Strupish, S.R.Nahorski and B.V.L.Potter, *Biochem.Soc.Trans.*, 1989, **16**, 992.

38) D.Lampe and B.V.L.Potter, *J.Chem.Soc.Chem.Commun.*, 1990, 1500.

39) C.E.Dreef, G.A. van der Marel and J.H. van Boom, *Recl.Trav.Chim.Pays-Bas*, 1987, **106**, 512.

40) J.R.Falck, A.Abdali and S.J.Wittenberger, *J.Chem.Soc.Chem.Commun.*, 1990, 953.

41) C.E.Dreef, W.Schiebier, G.A. van der Marel and J.H. van Boom, *Tetrahedron Lett.*, 1991, **32**, 6021.

42) R.F.Irvine, K.D.Brown and M.Berridge, *Biochem.J.*, 1984, **221**, 269.

43) V.Henne, G.W.Mayr, B.Grawowski, B.Koppitz and H.-D.Soling, *Eur.J.Biochem.*, 1988, **174**, 95.

44) R.Schafer, M.Nehls-Sahabandu, B.Grabowski, M.Dehlinger-Kremer and G.W.Mayr, *Biochem.J.*, 1990, **272**, 817.

45) G.D.Prestwich, J.Marecek, R.J.Mourey, A.B.Theibert, C.D.Ferris, S.K.Danoff and S.H.Snyder, *J.Am.Chem.Soc*, 1991, **113**, 1822.

46) A.M.Gurney and H.A.Lester, *Physiol.Rev.*, 1987, **67**, 583; J.A.McCray and D.R.Trentham, *Ann.Rev.Biophys.Biophys.Chem.*, 1989, **18**, 239.

47) J.W.Walker, J.Feeney and D.R.Trentham, *Biochemistry*, 1989, **28**, 3272.

48) J.W.Walker, A.V.Somlyo, Y.E.Goldmam, A.P.Sommlyo and D.R.Trentham, *Nature (London)*, 1987, **327**, 249.

49) M.Hirata, T.Sasaguri, T.Hamachi, T.Hashimoto, M.Kukita and T.Koga, *Nature (London)*, 1985, **317**, 723.

50) P.Westerduin, H.A.M.Willems and C.A.A. van Boekel, *Tetrahedron Lett.*, 1990, **31**, 6919.

51) M.F.Boehm and G.D.Prestwich, *Tetrahedron Lett.*, 1988, **29**, 5217.

52) J.F.Marecek and G.D.Prestwich, *Tetrahedron Lett.*, 1989, **30**, 5401.

53) S.T.Safrany, D.Sawyer, R.J.H.Wojcikiewicz, B.V.L.Potter and S.R.Nahorski, *FEBS Lett.*, 1990, **276**, 91.

54) M.Hirata, Y.Watanabe, T.Ishimatsu, T.Ikebe, Y.Kimura, K.Yamaguchi, S.Ozaki and T.Koga, *J.Biol.Chem.*, 1989, **264**, 20303.

55) M.Hirata, F.Yanaga, T.Koga, T.Ogasawara, Y.Watanabe and S.Ozaki, *J.Biol.Chem.*, 1990, **265**, 8404.

56) M.Hirata, Y.Watanabe, T.Ishimatsu, F.Yanaga, T.Koga and S.Ozaki, *Biochem. Biophys.Res.Commun.*, 1990, **168**, 379.

57) Y.Watanabe, M.Hirata, T.Ogasawara, T.Koga and S.Ozaki, *Bioorg.Med.Chem.Lett.*, 1991, **1**, 399.

58) H.A.J.Carlless and K.Busia, *Tetrahedron Lett.*, 1990, **31**, 1617.

59) C.Liu, B.V.L.Potter and S.R.Nahorski, *J.Chem.Soc.Chem.Commun.*, 1991, 1014.

60) M.J.Seewald, I.A.Aksoy, G.Powis, A.H.Fauq and A.P.Kozikowski, *J.Chem.Soc.*

Chem.Commun., 1990, 1638.

61)   A.P.Kozikowski, A.H.Fauq, I.A.Aksoy, M.J.Seewald and G.Powis, *J.Am.Chem.Soc*, 1990, **112**, 7403.

62)   A.P.Kozikowski, A.H.Fauq and A.H.Rusnak, *Tetrahedron Lett.*, 1989, **30**, 3365.

63)   S.V.Ley and F.Sternfeld, *Tetrahedron Lett.*, 1988, **29**, 5305.

64)   S.V.Ley, M.Parra, A.J.Redgrave, F.Sternfeld and A.Vidal, *Tetrahedron Lett.*, 1989, **30**, 3557.

65)   S.V.Ley, M.Parra, A.J.Redgrave and F.Sternfeld, *Tetrahedron*, 1990, **46**, 4995.

66)   S.T.Safrany, R.J.H.Wojcikiewicz, J.Strupish, S.R.Nahorski, D.Dubreuil, J.Cleophax, S.D.Gero and B.V.L.Potter, *FEBS Lett.*, 1991, **278**, 252.

67)   M.A.Polokoff, G.H.Bencen, J.P.Vacca, S.J.deSolms, S.D.Young and J.R.Huff, *J.Biol.Chem*, 1988, **263**, 11922.

68)   T.M.Connolly, V.S.Bansal, T.E.Bross, R.F.Irvine and P.W.Majerus, *J.Biol.Chem.*, 1987, **262**, 2146.

69)   C.Schultz, G.Gebauer, T.Metschies, L.Rensing and B.Jastorff, *Biochem. Biophys.Res.Comm.*, 1990, **166**, 1319.

70)   G.M.Burgess, R.F.Irvine, M.J.Berridge, J.S.McKinney and J.W.Putney Jr., *Biochem.J.*, 1984, **224**, 741.

71)   A.L.Willcocks, J.Strupish, R.F.Irvine and S.R.Nahorski, *Biochem.J.*, 1989, **257**, 297.

72)   G.V.Denis and C.E.Ballou, *Cell Calcium*, 1991, **12**, 395.

73)   T.McInally, I.Millichip, A.C.Tinker, C.Hallam, A.Kirk and A.Wallace, Two posters presented at the *11th International Symposium on Medicinal Chemistry*, (EFMC) Jerusalem, 1990; Copies available from T.McInally, Department of Medicinal Chemistry, Fisons plc, Pharmaceutical Division, Research and Development Laboratories, Bakewell Road, Loughborough, Leicestershire, LE11 0RH, United Kingdom.

74)   V.A.Estevez and G.D.Prestwich, *Tetrahedron Lett.*, 1991, **32**, 1623.

75)   S.R.Nahorski and B.V.L.Potter, *Trends Pharmacol. Sci.*, 1989, **10**, 139.

Chapter 9

Late Entries

9.1 Resolutions

A number of new resolution methods have been described for inositol derivatives.

A valuable enzymic resolution method involves the enantioselective acylation of racemic alcohols using lipase enzymes in organic solvents.[1] Application of this methodology to the racemic inositol bisacetals (1) and (4) allows effective resolution of these intermediates to be achieved.[2] For example, the lipase "amano lipase AY" acylates selectively the 4-position of

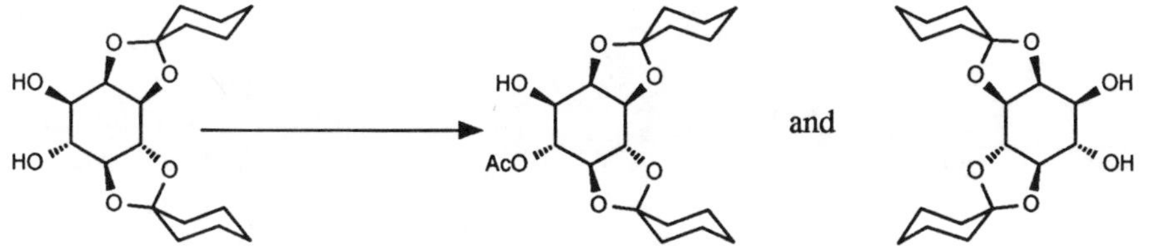

rac (1)   D-(2)   L-(3)

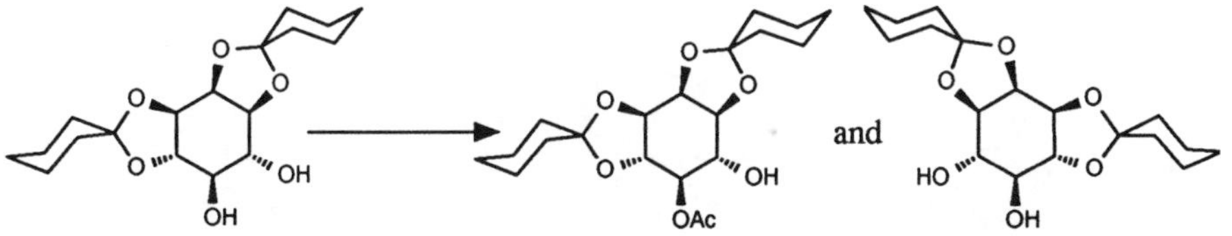

rac (4)   D-(5)   L-(6)

Fig 9 - 1

the D-enantiomer of racemic (1). By careful choice of conditions, the acetal (1) may be converted into a readily separable mixture of the optically active monoacetate D-(2), and the optically active unreacted bisacetal L-(3), in 100% and 98% enantiomeric excess respectively, *Fig 9-1*. In a similar fashion, the enzyme selectively acylates the 5-position of the D-enantiomer of racemic bisacetal (4), to give a mixture of the optically active monoacetate D-(5), and the optically active unreacted bisacetal L-(6) in 96% and 100% enantiomeric excess respectively. Hydrolysis of the acetate groups then provides access to both enantiomers of (1) and (4) in good optical purity.

Exploitation of the higher reactivity of the hydroxyl group at the 1-position of protected inositol derivatives, combined with the use of an optically active acylating agent (8) prepared from (R)-(-)- mandelic acid (7) as shown in *Fig 9-2*, leads to a flexible new route to optically active protected inositols.[3] Treatment of the 1,2-cyclohexylidine acetal (9) with the acylating

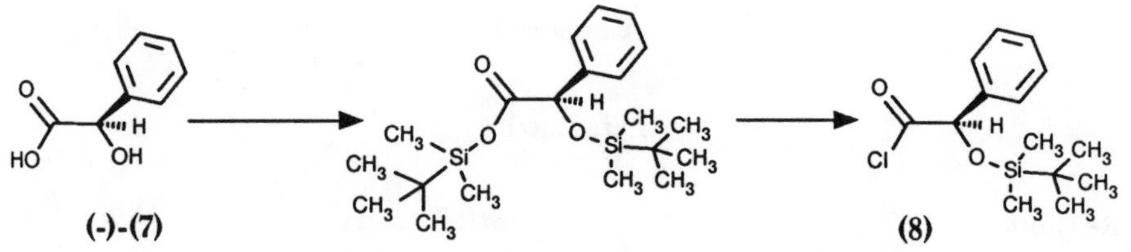

Fig 9 - 2

agent (8) resulted in selective acylation of the hydroxyl group at the 1-position, giving a
mixture of the diastereomeric esters (10) and (11) in 55% isolated yield, *Fig 9-3*. Direct
chromatographic separation of (10) and (11) proved impossible, but precipitation from
ether-hexane gave (11) in 96% diastereomeric excess. Treatment of the mixture of (10) and

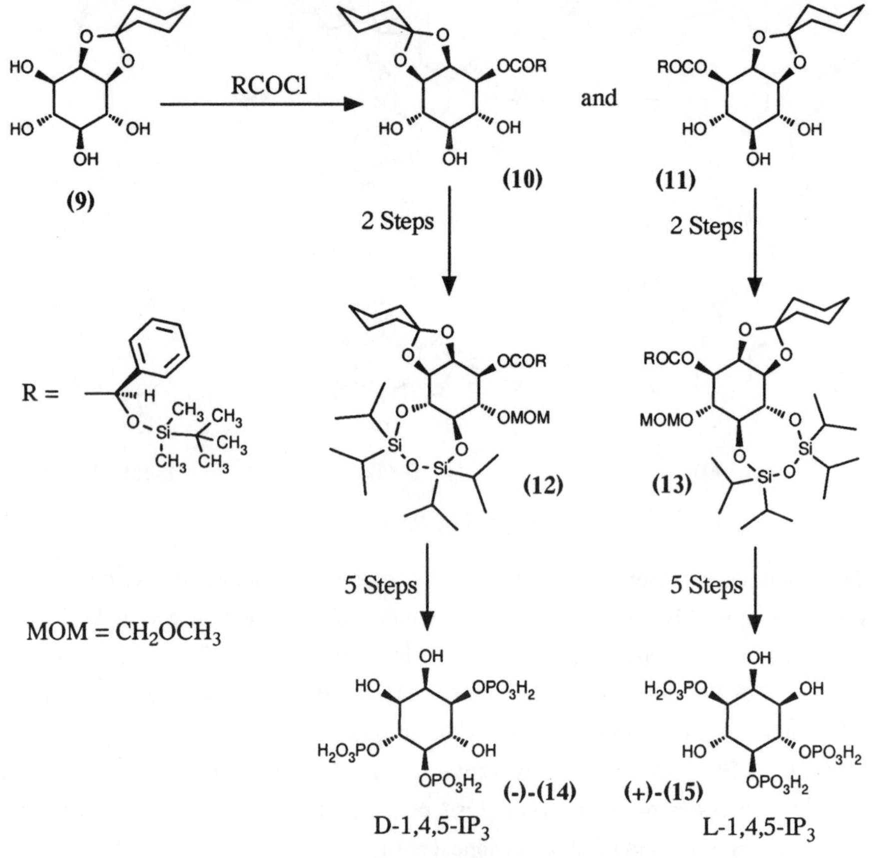

Fig 9 - 3

(11) with TIPS chloride gave selectively the 4,5-silylated products, which on conversion to a
mixture of the fully protected methoxymethyl ethers (12) and (13) proved separable by
normal chromatographic methods. Removal of the silyl and acyl protecting groups from (12)

and (13), followed by phosphorylation and deprotection then gave the enantiomers of 1,4,5-IP$_3$, (-)-(14) and (+)-(15). The optical rotations of these materials allowed the absolute stereochemistry of the triols (10) and (11), and the protected inositols derived from them, to be assigned.

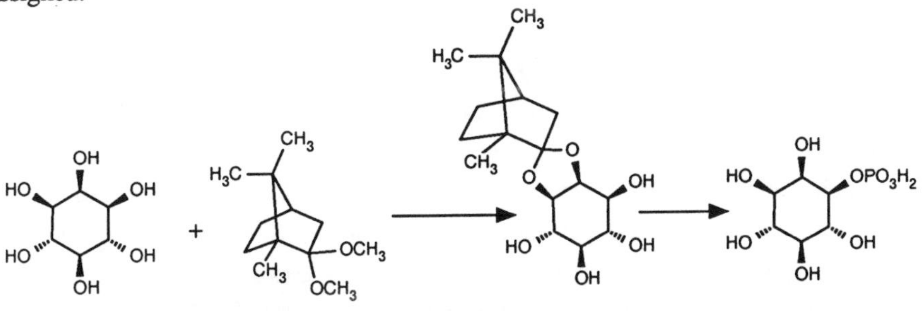

D-(16) D-(17) D-(18)

Fig 9 - 4

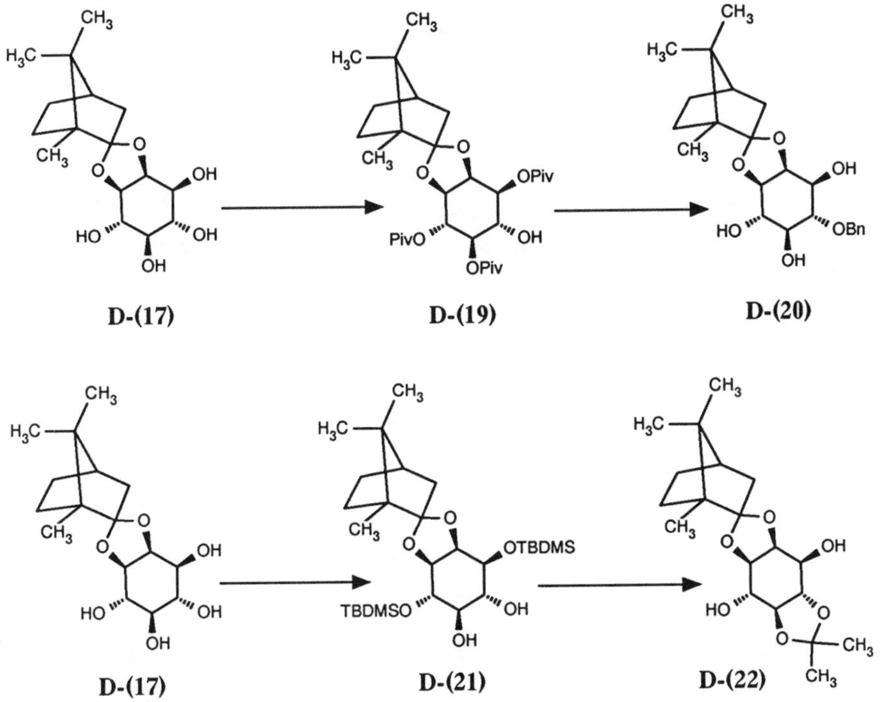

D-(17) D-(19) D-(20)

D-(17) D-(21) D-(22)

Fig 9 - 5

Reaction of inositol with the dimethyl acetal of D-camphor (16), gives rise to a complex mixture of mono and bisacetals, which may be partially hydrolysed to a separable mixture of the four possible *cis*-monoacetals.[4] Under carefully controlled hydrolytic conditions[5] a precipitation-driven equilibration of the original complex mixture occurs, resulting in recovery of the optically active D-monoacetal (17), in 70% yield and high optical purity, *Fig 9-4*. The tetrol (17) may be selectively phosphorylated at the 1-position with dibenzyl chlorophosphate, to give the protected phosphate in modest yield, which on deprotection gives D-inositol-1-phosphate, D-(18).[5] The acetal D-(17) is also useful for the synthesis of inositol polyphosphates. Reaction of D-(17) with 5 equiv. of pivaloyl chloride gives the 1,4,5-trispivaloyl derivative D-(19) in 48% isolated yield, *Fig 9-5*, which on benzylation and removal of the pivaloyl groups gives the optically active triol D-(20), which can be used for the preparation of D-1,4,5-IP$_3$.[6] Treatment of D-(17) with 2.2 equiv. of *t*-butyldimethylsilyl chloride results in the selective silylation of the 1 and 4-positions giving D-(21) in 50% isolated yield. Acetal formation, followed by removal of the two silyl groups then gives the optically active diol D-(22), a suitable intermediate for the synthesis of D-1,4-IP$_2$.

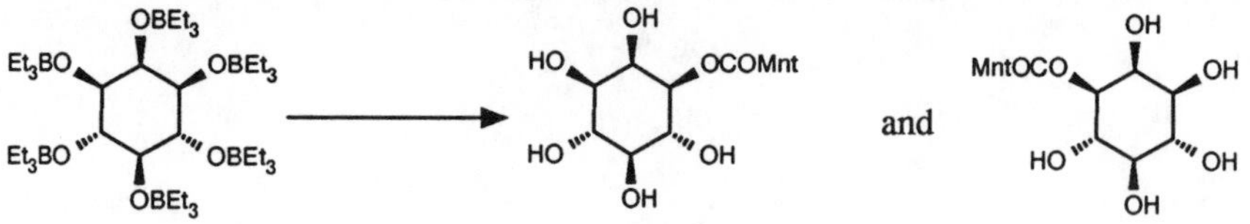

(23)   D-(24)   and   L-(25)

Fig 9 - 6

An extension of the selective reactions of partially stannylated inositols[7] (outlined in *Fig 3-13*) allows the direct conversion of the hexaboryl inositol (23) into a mixture of the diastereomeric monomenthoxy acetates D-(24) and L-(25), *Fig 9-6*, from which the individual diastereomers may be isolated in modest yield but having 98% diastereomeric

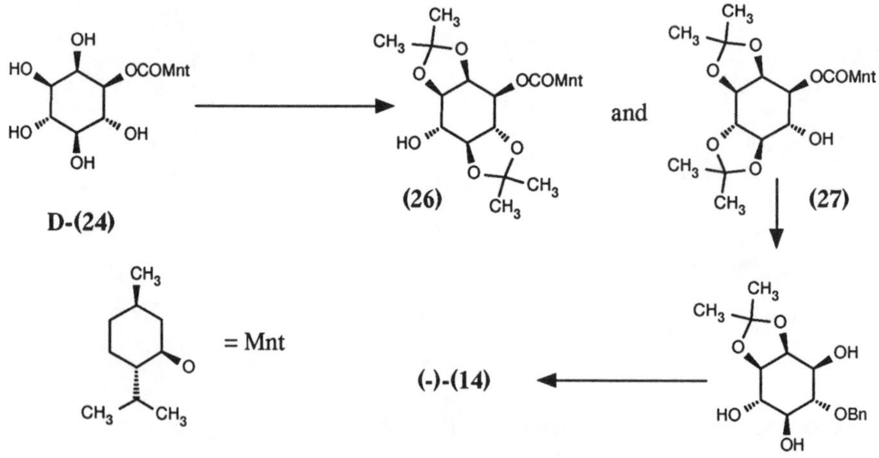

D-(24)   (26)   and   (27)

= Mnt

(-)-(14)   (+)-(28)

Fig 9 - 7

excess.[8] Treatment of D-(24) with 2,2-dimethoxypropane gives a mixture of the two possible bisacetals (26) and (27), *Fig 9-7*. Benzylation of (27), followed by selective hydrolysis of the less stable *trans* acetal group and hydrolysis of the menthoxyacetate ester then gives the optically active triol (+)-(28), which on phosphorylation and deprotection gives D-1,4,5-IP$_3$, (-)-(14).

9.2 Synthesis of 1,3,4-IP$_3$ and 1,3,4,5-IP$_4$

The optically active bisacetal (-)-(29), obtained by enzymic hydrolysis[9] of the corresponding bisacetate with cholesterol esterase, (outlined in *Fig 3-18*) has been used as a starting material for the synthesis of optically active 1,3,4-IP$_3$ and 1,3,4,5-IP$_4$.[10] Allylation of the diol (-)-(29), followed by hydrolysis of the less stable *trans* acetal gave the diol (+)-(30), *Fig 9-8*. Benzylation of (+)-(30), followed by hydrolysis of the *cis* acetal then gave the diol (+)-(31). Selective allylation at the more reactive 1-position of (+)-(31) gave (+)-(32), which on benzylation of the remaining free hydroxyl group and subsequent removal of the three allyl groups gave the triol (-)-(33). Phosphorylation and deprotection of (-)-(33) then gave D-1,3,4-IP$_3$.

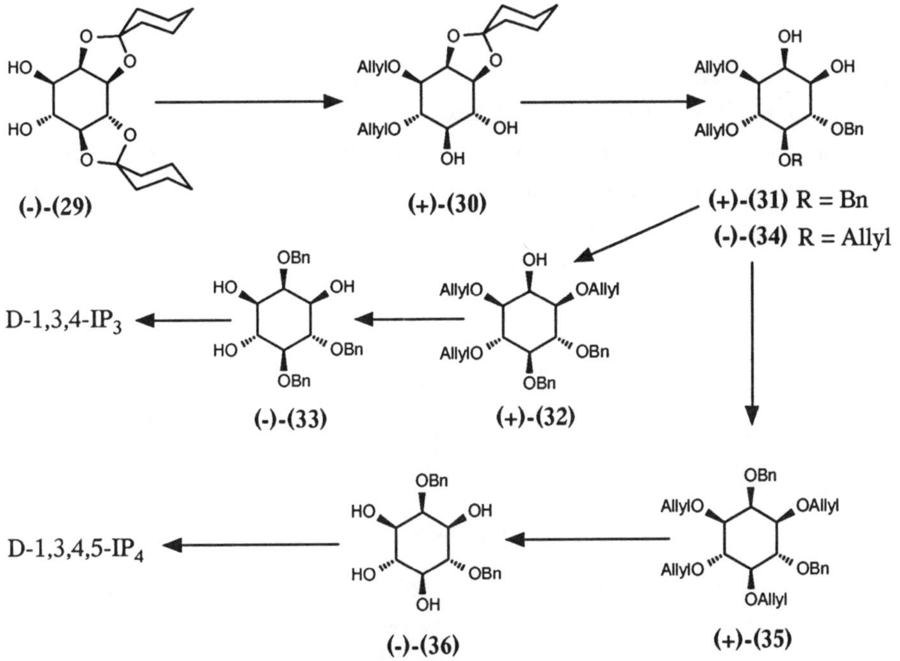

Fig 9 - 8

Alternatively, selective benzylation of the 6-position of diol (-)-(30), followed by allylation of the remaining hydroxyl group, and hydrolysis of the *cis* acetal gave the diol (-)-(34).

Selective allylation at the 1-position and subsequent benzylation as before then gave the fully protected inositol (+)-(35). Removal of the four allyl groups then gave the tetrol (-)-(36), which on phosphorylation and deprotection gave D-1,3,4,5-IP$_3$.

9.3 Analogues

Using methyl phosphinate, and protected inositols previously described, the inositol phosphate analogues (37), (38) and (39) have been prepared, *Fig 9-9*.[11] For all of these molecules the replacement of phosphate by the methylphosphonate group(s) led to compounds which were biologically inactive. The previously reported[12] trissulphate and trissulphonamide derivatives of 1,4,5-IP$_3$ (see Chapter 8) have now also been reported to be biologically inactive.

(37)   (38)   (39)

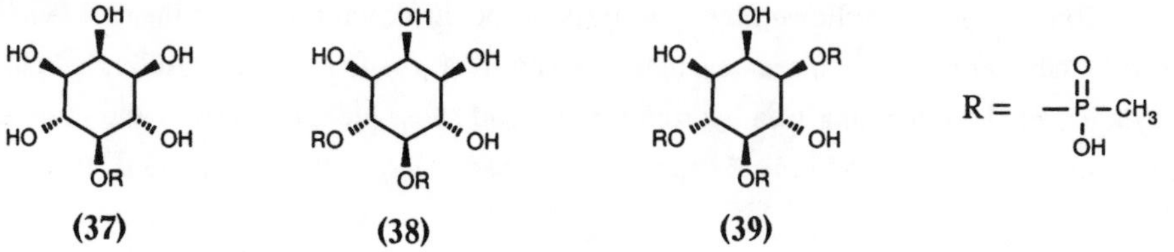

Fig 9 - 9

The synthesis of 3-azido-3-deoxy-inositol 2,4,5-trisphosphate has been reported,[13] using the azido-inositol (40)[14] as a starting material. Treatment of (40) with 1.2 equiv. of TIPS chloride led to selective formation of the triol (41) in 95% yield, *Fig 9-10*. Phosphorylation of (41) with tetrabenzylpyrophosphate, and subsequent deprotection with TMS bromide then gave the 3-azido analogue of 2,4,5-IP$_3$, *rac* (42). Preliminary biological results indicated that (42) has little biological activity as a promoter of calcium release.

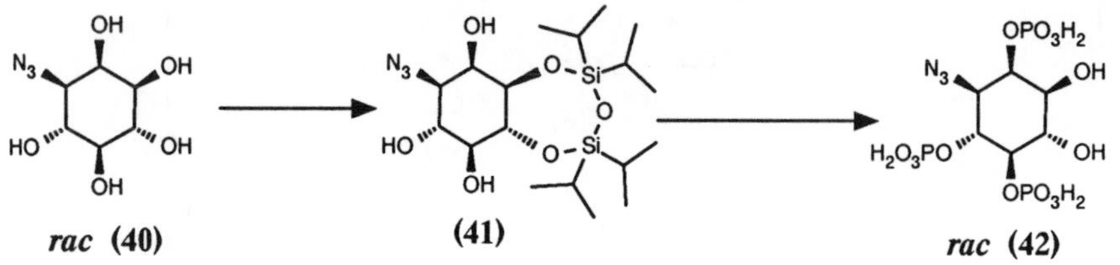

rac (40)   (41)   *rac* (42)

Fig 9 - 10

The synthesis of the individual enantiomers of 2,2-difluoro-2-deoxy-inositol 1,4,5-tris-phosphate has now been achieved, involving the resolution of the difluoro alcohol (43) *via* conversion to its diastereomeric camphanate esters.[15] After separation of the camphanates, and hydrolysis of the camphanate esters, hydrolysis of the acetal group of each of the enantiomers of (43), followed by phosphorylation and deprotection gave D-(44) and L-(45), *Fig 9-11*. D-(44) proved to be an excellent analogue of 1,4,5-IP$_3$, being a full agonist in

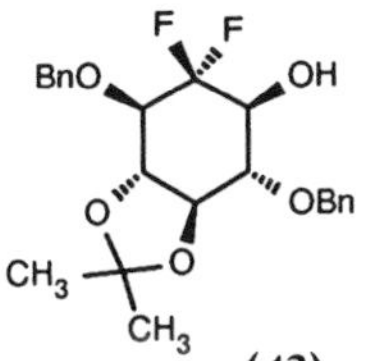

Fig 9 - 11

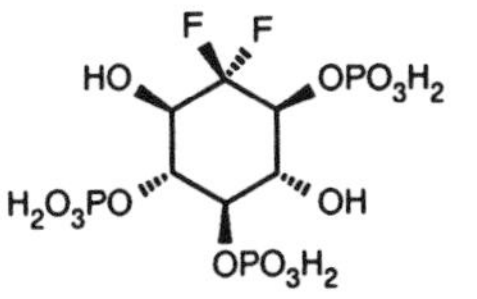

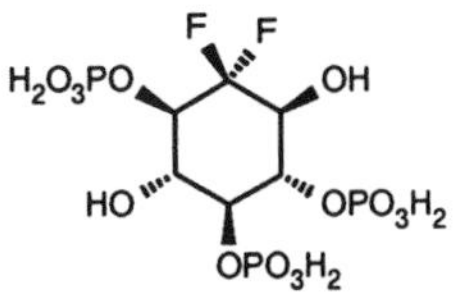

calcium release studies, and substrate for both the 5-phosphatase and 3-kinase enzymes which metabolise 1,4,5-IP$_3$. In stark contrast, L-(45) was not active in calcium release studies, and was not a substrate for the 5-phosphatase. L-(45) was however a potent inhibitor of both the 5-phosphatase and 3-kinase enzymes.

References

1)   A.M.Klibanov, *Acc.Chem.Res.*, 1990, **23**, 114.

2)   L.Ling, Y.Watanabe, T.Akiyama and S.Ozaki, *Tetrahedron Lett.*, 1992, **33**, 1911.

3)   K.S.Bruzik, J.Meyers and M.-D. Tsai, *Tetrahedron Lett.*, 1992, **33**, 1009.

4)   K.S.Bruzik and G.M.Salamonczyk, *Carbohydrate Res.*, 1989, **195**, 67.

5)   G.M.Salamonczyk and K.M.Pietrusiewicz, *Tetrahedron Lett.*, 1991, **32**, 4031.

6)   G.M.Salamonczyk and K.M.Pietrusiewicz, *Tetrahedron Lett.*, 1991, **32**, 6167.

7)   A.Zapata, R.Fernandez de la Pradilla, M.Martin-Lomas and S.Penades, *J.Org.Chem.*, 1991, **56**, 444.

8)   A.Aguilo, M.Martin-Lomas and S.Penades, *Tetrahedron Lett.*, 1992, **33**, 401.

9)   Y.-C.Liu and C.-S.Chen, *Tetrahedron Lett.*, 1989, **30**, 1617.

10) D.-M.Gou and C.-S.Chen, *Tetrahedron Lett.*, 1992, **33**, 721.

11) H.A.M.Willems, G.H.Veeneman and P.Westerduin, *Tetrahedron Lett.*, 1992, **33**, 2075.

12) P.Westerduin, H.A.M.Willems and C.A.A. van Boekel, *Tetrahedron Lett.*, 1990, **31**, 6919.

13) A.P.Kozikowski, A.H.Fauq, G.Powis, P.Kurian and F.T.Crews, *J.Chem.Soc. Chem.Commun.*, 1992, 362.

14) A.P.Kozikowski, A.H.Fauq, G.Powis, M.J.Seewald and I.A.Aksoy, *J.Am.Chem.Soc*, 1990, **112**, 7403.

15) D.Sawyer and B.V.L.Potter, *Bioorg.Med.Chem.Lett.*, 1991, **1**, 705.

Subject Index